REVISE OCR AS/A LEVEL
Biology A
REVISION GUIDE

Series Consultant: Harry Smith

Authors: Kayan Parker and Colin Pearson

Our revision resources are the smart choice for those revising for OCR AS/A Level Biology A. This book will help you to:

- **Organise** your revision with the one-topic-per-page format
- **Speed up** your revision with summary notes in short, memorable chunks
- **Track** your revision progress with at-a-glance check boxes
- **Check** your understanding with worked examples
- **Develop** your exam technique with exam-style practice questions and full answers.

Revision is more than just this Guide!

Make sure that you have practised every topic covered in this book, with the accompanying OCR AS/A Level Biology A Revision Workbook. It gives you:

- More exam-style practice and a 1-to-1 page match with this Revision Guide
- Guided questions to help build your confidence
- Hints to support your revision and practice.

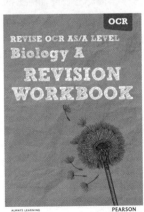

For the full range of Pearson revision titles across GCSE, BTEC and AS/A Level visit:

www.pearsonschools.co.uk/revise

PEARSON

Contents

MODULE 1
Practical skills are assessed throughout the course, both in written examination and in practical endorsement. As such, you will find notes on this throughout the guide.

A small bit of small print
OCR publishes Sample Assessment Material and the Specification on its website. This is the official content and this book should be used in conjunction with it. The questions in *Now try this* have been written to help you practise every topic in the book. Remember: the real exam questions may not look like this.

Using a light microscope

Light microscopes can be used to view **living** cells and tissues.

Advantages and disadvantages

✓ You can view the specimen in colour.

✓ You can view the specimen in real time.

✗ The light microscope has a low magnification (×1500) and a low resolution (200 nm).

✗ It is difficult to see the organelles of the cell.

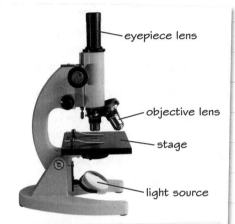
eyepiece lens
objective lens
stage
light source

🖩 Maths skills Calculating magnification

The magnification of a light microscope is the magnification of the eyepiece lens multiplied by the magnification of the objective lens.

For example:

Eyepiece lens = ×10

Objective lens = ×40

Magnification = 10 × 40 = ×400

You do not need to remember all of these labels, but you do need to know how the light microscope is used and the types of images you can see with it.

Cells as viewed under a light microscope

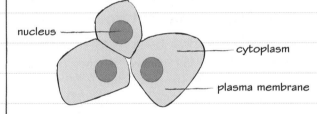
nucleus
cytoplasm
plasma membrane

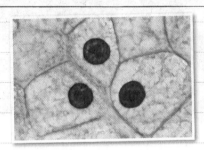

Worked example

Give two uses of a light microscope.   **(2 marks)**

Observe living cells and tissues.

Observe cellular events, such as mitosis, in real time.

Even at maximum magnification, it is only possible to observe the larger features of the cell, such as the nucleus or chromosomes during mitosis.

In order to see the organelles of the cell, you would need to use a transmission electron microscope (see page 2 for information about other microscopes).

Now try this

1 A cell is observed under an objective lens of ×20 and an eyepiece lens of ×10. What is the magnification of the cell?   **(1 mark)**

2 A scientist observes blood under a microscope. Give two advantages of using a light microscope to observe the blood cells.   **(2 marks)**

Make sure your answer is **specific** to the question. You should discuss the advantages of the light microscope in relation to the cells of the blood.

Using other microscopes

Electron microscopes

Transmission electron microscopes use electrons to produce 2D images of the inside of a cell. They have a maximum magnification of ×2 000 000.

Scanning electron microscopes use electrons to produce 3D images of the outside surface of cells. They have a maximum magnification of ×200 000.

Advantages and disadvantages

✓ Electron microscopes have a high magnification and a high resolution (0.1 nm).

✗ You can only view dead specimens.

✗ Specimens can be viewed only in black and white.

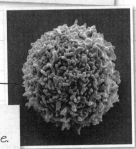

Plant root tip cell as viewed by a transmission electron microscope

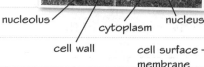

nucleolus   cytoplasm   nucleus

cell wall   cell surface membrane

Lymphocyte as viewed by a scanning electron microscope.

Laser scanning confocal microscope

This type of microscope uses lasers to produce images at different depths within the cell.

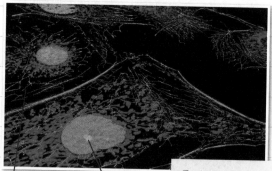

cytoskeleton   nucleus

Endothelial cells as viewed by a laser scanning confocal microscope.

Advantages and disadvantages

✓ You can view living cells in colour.

✓ You can observe cells in real time.

✓ Sections can be made through the cell either vertically or horizontally.

✗ Magnification and resolution are low compared to electron microscopes.

The difference between magnification and resolution

Magnification is the number of times greater the image is than the specimen.

Resolution is the ability to distinguish between two points on an image. The higher the resolution, the sharper the image.

Worked example

Describe and explain which microscope you would use to observe the detailed ultrastructure of a cell.   **(3 marks)**

I would use a transmission electron microscope because this type of microscope has a high magnification of ×2 000 000 and a high resolution of 0.1 nm.

Now try this

Three types of microscope are listed in the table below. Put a tick in each column that describes a function of that microscope.

Type of microscope	Can observe whole cells and tissues	Can observe organelles	Can observe cell surfaces	Can observe a certain depth within a cell
Transmission electron microscope				
Scanning electron microscope				
Laser scanning confocal microscope				

(3 marks)

Preparing microscope slides

🧪 **Practical skills**
How to prepare a slide for a light microscope

If you are preparing a slide for a specimen that is in solution, e.g. cheek cells in saliva, add a drop of specimen onto the slide and then add a drop of stain. You can then cover your specimen with a cover slip.

slowly lower

cover slip

water droplet

slide

1 Take a thin slice of your specimen and place it onto a clean microscope slide.

2 Add a drop of water or a drop of stain onto the specimen.

3 Take a clean cover slip and lower it slowly onto the specimen, taking care to avoid air bubbles.

How to use an eyepiece graticule

Line up the eyepiece graticule and the stage micrometer, using the ×4 objective lens. In this example, 80 small divisions on the eyepiece graticule correspond to 18 stage micrometer divisions. This represents

$18 \times 10\,\mu m = 180\,\mu m$

Therefore, there is a distance between each eyepiece graticule division of

$\frac{180}{80} = 2.25\,\mu m$

You can now use the eyepiece graticule to measure objects on the microscope stage.

eyepiece graticule scale

50   60   70
0 1 2 3 4 5 6 7 8 9 10

stage micrometer scale

The eyepiece graticule is divided into 100 divisions and has an arbitrary scale. The stage micrometer is 1mm in length and divided into 100 divisions, each one 10μm apart.

Uses of staining:

- To highlight a particular organelle, e.g. the nucleus.
- To distinguish between different cell types, e.g. red blood cells and white blood cells.
- To distinguish different tissue types, e.g. smooth muscle tissue.

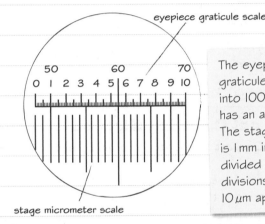

Red blood cell stained with eosin and white blood cells stained with haematoxylin.

Worked example

A scientist wants to observe onion cells with a light microscope. Describe how the microscope slide should be prepared.
(3 marks)

A thin section of the specimen is placed onto a microscope slide, covered with a drop of stain and then covered with a cover slip.

Now try this

🔢 **Maths skills**

1 The eyepiece graticule was lined up with the stage micrometer using the ×10 lens. This time, 100 divisions on the eyepiece graticule were lined up with 10 divisions on the stage micrometer. What is the distance between each eyepiece graticule division? **(2 marks)**

2 A blood smear is stained with eosin and haematoxylin. Explain what the observer should expect to see using a light microscope. **(2 marks)**

Calculating magnification

 Maths skills You need to know how to calculate the magnification of a cell or organelle, using the appropriate units of measurement. You should be able to use and manipulate the magnification formula.

How to work out the magnification

Photos of specimens taken down the microscope are called micrographs. It is easy to work out the magnification of the image if you know:

- the size of the image
- the actual size of the specimen.

The formula for this is:

$$\text{magnification} = \frac{\text{image size}}{\text{actual size}}$$

Rearranging the formula

You can work out how to rearrange the magnification formula using a formula triangle.

For example, if you want to work out the image size, cover the top part of the formula triangle. This leaves the 'actual size' and the 'magnification'. Therefore the formula is:

image size = actual size × magnification

> You can remember the formula triangle as the first letters spell out 'I AM'.

Working out the magnification of a cell

It is 40 mm. Convert the length into µm by multiplying the length in mm by 1000.

The actual size of the mitochondrion is 2 µm. Use the magnification formula to work out the magnification:

$$\frac{40\,000\,\mu m}{2\,\mu m} \times 20\,000$$

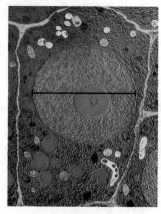

Worked example

Calculate the magnification of the microscope. The actual length of the mitochondrion is 1.9 µm and the image length is 38 mm. Show your working.

(2 marks)

Image size = 38 mm = 38 000 µm

$$\text{magnification} = \frac{38\,000\,\mu m}{1.9\,\mu m}$$

magnification = ×20 000

> Always remember to convert the image size into µm from the size in mm. For example, 10 mm = 10 000 µm.

Now try this

1 (a) How many µm in 12 cm? **(1 mark)**
  (b) How many µm in 16 mm? **(1 mark)**
2 Calculate the actual width of the plant cell. The magnification is ×15 000. Show your working. **(2 marks)**

Eukaryotic and prokaryotic cells

Eukaryotic cells have a nucleus inside a nuclear envelope, and other organelles that are surrounded by a membrane. Prokaryotic cells do not contain a nucleus.

Eukaryotic cells

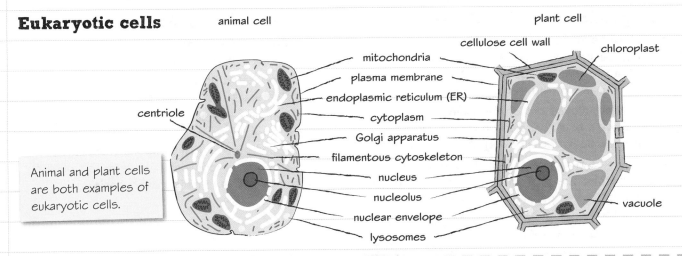

animal cell

plant cell

cellulose cell wall

chloroplast

mitochondria
plasma membrane
endoplasmic reticulum (ER)
cytoplasm
Golgi apparatus
filamentous cytoskeleton
nucleus
nucleolus
nuclear envelope
lysosomes

centriole

vacuole

Animal and plant cells are both examples of eukaryotic cells.

Prokaryotic cell

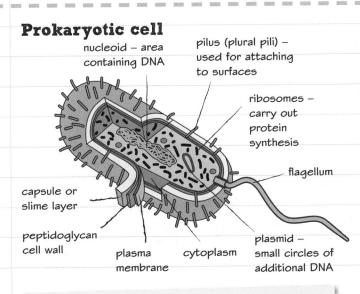

nucleoid – area containing DNA

pilus (plural pili) – used for attaching to surfaces

ribosomes – carry out protein synthesis

flagellum

capsule or slime layer

peptidoglycan cell wall

plasma membrane

cytoplasm

plasmid – small circles of additional DNA

Prokaryotic cells, for example bacteria, have no organelles. All of the functions usually carried out inside organelles are carried out in the cytoplasm.

Organelles and their functions

Organelle	Function
Nucleus	Contains the cell's DNA.
Nucleolus	Makes RNA and ribosomes.
Nuclear envelope	Surrounds the nucleus.
Mitochondria	Site of aerobic respiration.
Golgi body	Modifies proteins and packages them for transport.
Rough endoplasmic reticulum	Contains ribosomes. Transports proteins made by ribosomes.
Smooth endoplasmic reticulum	Lipid synthesis.
Ribosomes	Protein synthesis.
Chloroplasts	Site of photosynthesis.
Plasma membrane	Controls what enters and exits the cell.
Cellulose cell wall	Supports the cell.
Centrioles	Part of cell division.
Lysosomes	Break down waste materials in the cell.
Cilia	Movement of cells or substances.
Flagella	Movement of cells.

Worked example

State three differences between a eukaryotic cell and a prokaryotic cell.   **(3 marks)**

Eukaryotic cells have a nucleus.

Eukaryotic cells have organelles.

Prokaryotic cells have a peptidoglycan cell wall.

Now try this

1 Compare the movement of prokaryotic and eukaryotic cells.   **(2 marks)**
2 State the structural and functional differences between smooth and rough endoplastic reticulum.   **(4 marks)**

The secretion of proteins

Proteins are produced by the nucleus and ribosomes by the processes of transcription and translation. For more detail on these processes, see page 28.

The production of proteins

- Each gene in the nucleus codes for a particular protein. When the cell wants to make a protein, a messenger RNA (mRNA) copy is made of the gene.
- The mRNA moves out of the nucleus and travels to ribosomes on the rough endoplasmic reticulum (rough ER).
- Ribosomes translate the mRNA and produce the protein. The protein travels through the rough ER and is packaged into a vesicle to travel to the Golgi body.

The secretion of proteins

- The Golgi body modifies the protein, usually by adding carbohydrate chains.
- The processed proteins are then packaged into vesicles and travel to the plasma membrane.
- The membrane of the vesicle is made of the same phospholipids as the plasma membrane. The vesicle fuses with the plasma membrane and the protein contents are released outside of the cell. This is called **exocytosis**.

The interrelationship between the organelles involved in the production and secretion of proteins

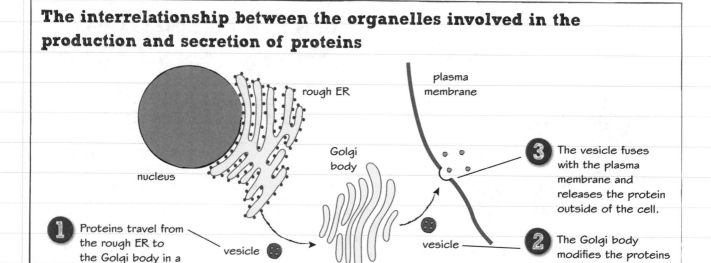

nucleus

rough ER

plasma membrane

Golgi body

3 The vesicle fuses with the plasma membrane and releases the protein outside of the cell.

1 Proteins travel from the rough ER to the Golgi body in a vesicle.

vesicle

vesicle

2 The Golgi body modifies the proteins and packages them into a vesicle.

Worked example

What is the role of the Golgi body in the secretion of proteins? **(2 marks)**

Modifies the protein.

Packages the protein into vesicles.

The Golgi body can also be called the Golgi apparatus or the Golgi complex.

Make sure you use the term **modifies** when describing the function of the Golgi body – not changes, alters or processes.

Now try this

1 Describe the role of the rough ER in protein secretion. **(3 marks)**

2 Plasma cells make large numbers of antibodies. Explain the relationship between ribosomes, the rough ER and the Golgi body in these cells. **(4 marks)**

 Remember, antibodies are proteins.

The cytoskeleton

You need to know about the importance of the cytoskeleton and its roles in cell stability and movement.

Structure

The cytoskeleton is a network of fibres that run throughout the cell. Like the body's skeleton, the cytoskeleton supports the cell and gives it shape and structure.

There are three different types of fibre in the cytoskeleton:

1 microfilaments

2 microtubules

3 intermediate filaments.

The cytoskeleton is also involved in transport and movement.

Transport

Proteins, called microtubule motors, move up and down the cytoskeleton, carrying vesicles full of other proteins to different areas of the cell, or for transport out of the cell.

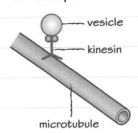

— vesicle
— kinesin
microtubule

Kinesin carrying a vesicle along a microtubule. Kinesin moves with a walking motion.

Cell movement

In eukaryotic cells, some microtubules extend out from the cell's plasma membrane, where they form flagella (undulipodia) or cilia. Both of these structures are made from nine microtubules arranged into a circle, with two central microtubules.

The 9 + 2 arrangement of microtubules in flagella and cilia.

Undulipodia and cilia in eukaryotes

kinesin

Undulipodia are used for transport of the whole cell, for example, sperm cells.

Cilia are used to move along substances outside of the cells. For example, cilia in the oviducts move the ovum towards the uterus.

Try to remember an example for each function of the cytoskeleton. For example, the trachea also contains cilia, to move mucus away from the lungs.

Worked example

Describe the role of the cytoskeleton. **(3 marks)**

The cytoskeleton gives the cell its structure. It also allows the transport of vesicles inside the cell, as microtubule motors move up and down the microtubules of the cytoskeleton. Some microtubules extend from the plasma membrane to form flagella / undulipodia and cilia to help cell movement.

Prokaryotic flagella are different

Flagella in eukaryotes are usually called undulipodia, whereas flagella in prokaryotes are always referred to as flagella.

Prokaryotic flagella are also made of microtubules, but be aware that these have a different internal structure (not the 9 + 2 arrangement).

Now try this

1 Describe how the cytoskeleton is used to transport proteins around the cell. **(3 marks)**
2 How are microtubules used in the process of fertilisation? **(4 marks)**

 Think about the sperm cells and the ovum cells.

The properties of water

Water is a common chemical but it is essential for life. Water is used inside the bodies of all organisms, and also forms the habitat for many species on Earth.

The structure of water

Water is made up of one oxygen atom and two hydrogen atoms. It is a dipolar molecule, which means that it contains a positive and negative charge. The oxygen atom has a slightly negative charge (shown by the symbol δ^-) and the hydrogen atoms have a slightly positive charge (δ^+).

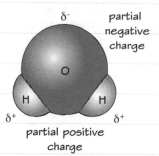

Hydrogen bonding in water

Hydrogen bonds can form between the oxygen atom of one water molecule and a hydrogen atom of another.
This helps to keep water molecules together (cohesion).

Water does not heat up or cool down easily. The molecules are held together quite tightly by hydrogen bonds. It takes energy to break these bonds, and so the high heat capacity provides thermal stability, making water a stable habitat.

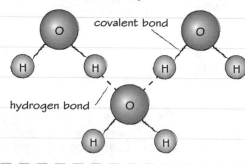

Properties of water

Property of water	Benefit to organisms
Thermal stability	Stable environment for aquatic organisms.
Forms a liquid	Transport medium in animals and plants.
Low density of ice	Ice floats on water and insulates the water below.
Surface tension	Small animals can move across the surface of the water.
Cohesion	Water molecules form a strong water column that moves up the xylem in plants.
Solvent	Chemicals dissolved in water in the cytoplasm can take part in reactions.
Metabolic	Water is used to break bonds in hydrolysis and make bonds in condensation reactions.
Transparency	Water allows light to move through it so aquatic plants can photosynthesise.

Worked example

How is water important for aquatic life?   **(4 marks)**

Water provides a stable environment for aquatic organisms because of its high specific heat capacity.

It is a liquid between 0°C and 100°C so it can be used as a transport medium in aquatic plants and animals.

Ice has a lower density than water, so it floats on top of the water, insulating the water below.

Water is transparent so light can reach aquatic plants for photosynthesis.

Aquatic means 'lives in water'.
This question is worth four marks so you must list four properties.
You can use any property of water so long as you can apply it to the question.

Now try this

1 How does the structure of water aid its function as a transport medium?   **(3 marks)**

2 Explain how water is used as a coolant in multicellular organisms.   **(3 marks)**

The biochemistry of life

Biological molecules are made from a small selection of elements, but sometimes use inorganic ions as part of a biological process.

Elements that make up biological molecules

Biological molecules	Elements
Carbohydrate	C, H, O
Lipid	C, H, O
Protein	C, H, O, N, S
Nucleic acids	C, H, O, N, P
Haemoglobin	As protein, Fe^{2+}
DNA polymerase	As protein, Zn^{2+}
Chlorophyll	C, H, O, N, Mg^{2+}

C = carbon, H = hydrogen, O = oxygen, N = nitrogen, S = sulfur, P = phosphorus, Fe^{2+} = iron ion, Zn^{2+} = zinc ion, Mg^{2+} = magnesium ion

Monomers and polymers

- A **monomer** is a single molecule, for example, a molecule of glucose.
- A **polymer** is a chain of monomers bonded together; for example, amylose is a polymer of glucose monomers.

Monomers bond together through a condensation reaction. A covalent bond forms between two adjacent monomers and a molecule of water is also produced.

Adding a molecule of water breaks covalent bonds. This is called a hydrolysis reaction.

The inorganic ions (e.g. Fe^{2+}, Zn^{2+}, Mg^{2+}) attached to biological molecules are known as prosthetic groups.

Condensation reaction

A covalent bond is formed between two monomers by the removal of OH from one monomer and H from the other.

monomers linked by covalent bond

Hydrolysis reaction

A covalent bond between two monomers is broken by the addition of a water molecule. OH is added to one monomer and H to the other.

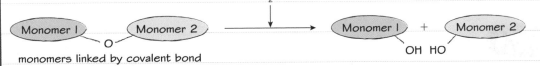

monomers linked by covalent bond

Worked example

Maltose is a disaccharide formed from two glucose molecules. Describe how a molecule of maltose is formed.　**(3 marks)**

A hydroxyl (OH) group from one glucose and a hydrogen atom (H) from the other glucose are removed and are joined to form a molecule of water.

A covalent bond is formed between the two molecules of glucose.

This is called a condensation reaction.

It is okay to refer to the hydroxyl group as OH and the hydrogen atom as H in your answer.

Make sure you use and spell key terms correctly. The key terms in this answer are **monomer**, **covalent** and **condensation reaction**.

Now try this

1　Which of the following formulae could be for a lipid?
　　A $C_6H_{12}O_6$　B $CH_3(CH_2)_{10}CO_2H$　C NH_2CH_2COOH　**(1 mark)**
2　Give two examples of a prosthetic group in biological molecules.　**(2 marks)**

Glucose

Glucose is an important molecule for life. It is oxidised in respiration to release energy and is used to form many disaccharides and polysaccharides.

The structure of glucose and ribose

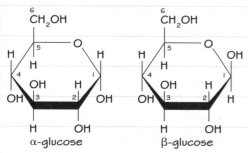

α-glucose

β-glucose

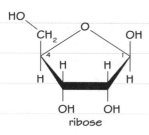

ribose

Disaccharides

A disaccharide is two monosaccharides bound together by a glycosidic bond. Most disaccharides contain at least one glucose monosaccharide.

Common disaccharides

Disaccharide	Monosaccharides
Maltose	glucose + glucose
Sucrose	glucose + fructose
Lactose	glucose + galactose

Glucose and ribose are both monosaccharides. Glucose is a hexose sugar (six carbon atoms) and ribose is a pentose sugar (five carbon atoms). Glucose comes in two forms: α-glucose (found in starch and glycogen) and β-glucose (found in cellulose).

Synthesis of glycosidic bonds

Glycosidic bonds form between two monosaccharides using a condensation reaction. The same mechanism is used to add a monosaccharide onto a polysaccharide.

In maltose, this bond forms between carbon 1 on the first α-glucose and carbon 4 on the second, so it is an α-1-4-glycosidic bond.

Breakdown of glycosidic bonds

A molecule of water is added to break the glycosidic bond. This is a hydrolysis reaction. The hydroxyl group (OH) bonds to carbon 1 on one glucose and the hydrogen atom (H) bonds to carbon 4 on the other glucose.

Worked example

Compare the structures of α-glucose and ribose.

(3 marks)

They are both monosaccharides.

They both form a ring structure.

Ribose has 5 carbons and glucose has 6 carbons.

The question is worth three marks, so make sure that you make three different points.

You may be asked to draw the structure of α- or β-glucose. β-glucose is the same as α-glucose, except the hydroxyl group (OH) and hydrogen atom (H) on carbon 1 are the opposite way around.

Now try this

1 Draw the process of two α-glucose molecules forming an α-1-4-glycosidic bond. **(2 marks)**
2 How do the glycosidic bonds differ in glycogen and cellulose? **(3 marks)**

Starch, glycogen and cellulose

Starch, glycogen and cellulose are important polysaccharides. They are all made of glucose but have very different structures.

Structure of starch

branched structure with few long branches

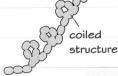
coiled structure

Amylose

Amylopectin

Starch is an energy storage molecule found in plants. It is the source of carbohydrate in our food.

Starch is made of two different polysaccharides – amylose and amylopectin. They are both made of α-glucose monomers but have different structures.

Structure of glycogen

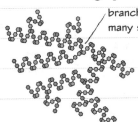
branched structure with many short branches

Glycogen is an energy storage molecule found in animals and people. Any excess glucose in our diet is stored as glycogen in the liver and muscles. When our blood glucose levels are low, glycogen is converted back into glucose.

Glycogen is made of α-glucose monomers and has a similar structure to amylopectin.

Structure of cellulose

Cellulose is found in the cell walls of plant cells and gives structure and support. It is made of β-glucose monomers that form long straight chains. Hydrogen bonds form between glucose molecules on adjacent strands to provide strength to the cells.

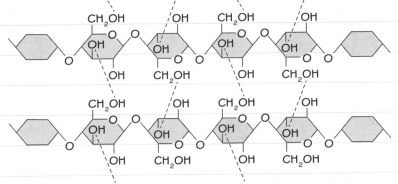

Properties and functions

Polysaccharide	Properties and functions
Starch	• Energy store in plants • Insoluble • Does not affect the water potential of the plant • Branches can be easily broken so that glucose can be used in respiration.
Glycogen	• Energy storage in animals • Insoluble • Does not affect the water potential of the animal cell • Branches can be easily broken so that glucose can be used in respiration.
Cellulose	• Structural carbohydrate in plant cell walls • Hydrogen bonds between cellulose chains give strength to cell walls.

Worked example

Compare the structure and function of starch and glycogen.   **(4 marks)**

Both are made from α-glucose.

Both are energy storage molecules.

Amylose and amylopectin are coiled but amylopectin has few long branches.

Glycogen is coiled and has many short branches.

Now try this

Describe the structure and function of cellulose.   **(3 marks)**

You should include the type of glucose, and how hydrogen bonding contributes to its function.

Triglycerides and phospholipids

Macromolecules are very large molecules, usually made of a long chain of the same monomers. Triglycerides and phospholipids are both examples of macromolecules that are made up of different types of monomer.

Tryglycerides

Triglycerides are a type of lipid consumed in our diet that we use as an energy source. The triglycerides are digested in our small intestines into fatty acids and glycerol.

These products are used by the body's cells to produce phospholipids – a part of a cell's plasma membrane. They can also be used to store fat in the body as triglycerides.

If the fatty acid chain contains no double bonds it is **saturated**.

If the fatty acid chain contains at least one double bond it is **unsaturated**.

Saturated fatty acids are more compact, due to the absence of any double bonds.

Unsaturated fatty acids with one double bond are called monounsaturated. Fatty acids with two or more double bonds are called polyunsaturated.

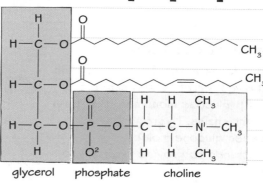

Structure of a triglyceride

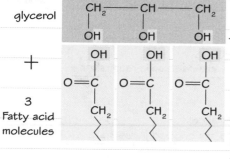

glycerol

Structure of a phospholipid

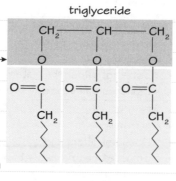

glycerol   phosphate   choline

Synthesis of triglycerides

glycerol

+

3 Fatty acid molecules

$3H_2O$

Condensation reaction forms three ester bonds between the fatty acids and glycerol.

triglyceride

Worked example

What is the difference between a saturated and an unsaturated fatty acid?   **(2 marks)**

In saturated fatty acids, all the possible bonds made by the carbon atoms are with hydrogen. Unsaturated fatty acids have a least one double bond.

Now try this

1 Contrast the structures of triglycerides and phospholipids.   **(2 marks)**
2 Draw diagrams to show how the ester bonds in a triglyceride are broken.   **(3 marks)**

Use of lipids in living organisms

Lipids are used in living organisms as an energy source, to produce phospholipids and cholesterol for use in the plasma membranes, and to make steroid hormones. Triglycerides (fats) are a subgroup of lipids.

Properties of lipids

Lipids are:

- large, organic molecules
- non-polar
- insoluble in water
- soluble in alcohol.

Lipids contain fatty acids, which are long chains of carbon and hydrogen.

Five uses of triglycerides

1 **Energy source:** triglycerides can be used by prokaryotes and eukaryotes in respiration to produce ATP.

2 **Energy storage:** mammals store triglycerides in adipose tissue.

3 **Protection:** body organs are surrounded by fat, which protects them from damage during sudden movement.

4 **Insulation:** Animals that live in cold climates have a layer of adipose tissue called 'blubber' that insulates them from the cold.

5 **Buoyancy:** Aquatic animals use their blubber to help them to float, as triglycerides are less dense than water.

Uses of phospholipids

Phospholipids are the main component of plasma membranes, where they form a **bilayer**.

This is because the phosphate head is **hydrophilic** and attracts water, whereas the lipid tail is **hydrophobic** and repels water.

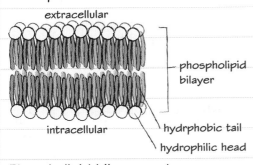

extracellular

phospholipid bilayer

intracellular

hydrphobic tail

hydrophilic head

Phospholipid bilayers act as a partially permeable membrane, allowing only small or non-polar molecules into the cell.

Uses of cholesterol and steroids

Cholesterol and steroids have a very similar structure.

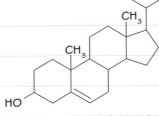

cholesterol

Cholesterol adds stability to plasma membranes. Animals that live in cold regions have more cholesterol in their membranes to prevent them from freezing.

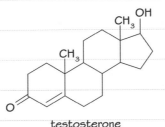

testosterone

Steroids are hormones such as oestrogen and testosterone. Although they are large, because they are non-polar they can cross the plasma membrane easily.

Worked example

How does the structure of triglyceride make it a useful molecule for energy storage?   **(2 marks)**

Triglycerides release double the amount of energy per gram, compared to glucose. This is due to triglycerides containing more hydrogen and less oxygen than glucose.
Triglycerides are insoluble in water, so they do not alter the water potential of the cell.

Carbohydrates release 17 kJ per gram whereas lipids release 37 kJ per gram.

Remember, water potential is more negative when solutes are dissolved in water.

Now try this

1 List the uses of lipids in Arctic seals.   **(3 marks)**
2 Explain why lipid-based hormones are faster acting than protein-based hormones.   **(2 marks)**

Amino acids

Amino acids are the molecules that make up proteins.

There are 20 different amino acids that make up proteins and all of them have a similar structure.

Amino acids are all made out of carbon, hydrogen, oxygen and nitrogen, but some also contain sulfur.

General structure of an amino acid

amino group   carboxyl group

The R group is different in each type of amino acid.

Formation and break down of peptide bonds

Two amino acids can join together with a **peptide bond** to form a **dipeptide**. Many amino acids joined together are called a **polypeptide**. This can then fold into a protein.

A hydroxyl group from the carboxyl group on one amino acid and a hydrogen from the amino group on the other amino acid are removed and form a water molecule. A peptide bond forms between the carboxyl and amino groups. The amino acids are now a dipeptide. This is a **condensation** reaction.

The peptide bond can be broken by the addition of a water molecule. The hydroxyl group is added to the carboxyl group of the first amino acid and the hydrogen is added to the amino group of the second amino acid. This is a **hydrolysis** reaction.

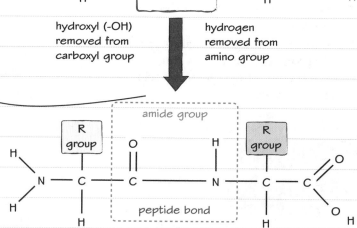

hydroxyl (-OH) removed from carboxyl group

hydrogen removed from amino group

amide group

peptide bond

A condensation reaction between two amino acids.

water

Worked example

Compare the formation of glycosidic bonds and peptide bonds.   **(3 marks)**

Both are condensation reactions.

Both release a molecule of water.

Glycosidic bonds are between two glucose molecules and peptide bonds are between two amino acids.

All bonds formed between organic molecules are condensation reactions that release a molecule of water. Remember:

Glucose → Glycosidic bond
Lipid → Ester bond
Protein → Peptide bond

Now try this

1 Describe the general structure of an amino acid.
    (3 marks)

2 Hydrogen bonds often form between adjacent amino acids in a protein. Explain why.   **(2 marks)**

Protein

The sequence of amino acids in a protein is its primary structure. Proteins fold into a secondary and then tertiary structure. Some proteins also have a quaternary structure.

Primary structure

Amino acids joined together by peptide bonds are called a **polypeptide**. Each protein is different because the amino acid sequence is different in each polypeptide.

Some amino acids are polar and some are non-polar. Other amino acids have a positive or negative charge. Amino acids can also vary in size. All of this determines how a polypeptide will fold.

Secondary structure

The polypeptide chain folds into one of two structures; an α-helix or a β-pleated sheet.

Both of these structures are held together by **hydrogen bonds**. This type of bond forms between the amino group (NH_2) of one amino acid and the carboxyl group (C=O) of another amino acid further down the chain.

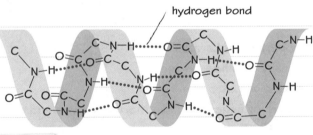

α-helix

β-sheet
(pleated sheet)

Tertiary structure

The α-helices and β-pleated sheets, along with areas of straight chains of amino acids, can fold further to form a complex three-dimensional structure. This is held together by:

- hydrogen bonds (C=O – – – N–H)
- ionic bonds (R^+ – – – R^-)
- disulfide bridges (–S–S–)
- hydrophobic and hydrophilic interactions.

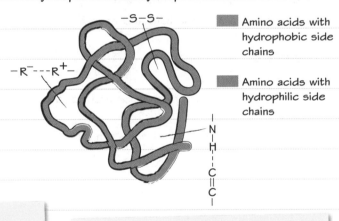

Amino acids with hydrophobic side chains

Amino acids with hydrophilic side chains

Note how the hydrophobic portions are in the centre of the protein and the hydrophilic portions are on the outside.

Quaternary structure

Some proteins, such as haemoglobin, form a quaternary structure. This is made of more than one polypeptide chain. The structure of haemoglobin is shown on page 17.

The structure of haemoglobin is shown on page 17.

Worked example

Explain what gives a protein its tertiary structure.
(4 marks)

The sequence of amino acids in the polypeptide.
Charged amino acids forming ionic bonds.
Polar and non-polar amino acids causing hydrophilic and hydrophobic interactions.
Hydrogen bonding between amino acids.

Only amino acids that contain sulfur, like cysteine, can form disulfide bridges.

Now try this

1 Haemoglobin contains four polypeptide chains. What type of structure is this? **(1 mark)**
2 Where in a protein would you expect to find non-polar amino acids? **(1 mark)**

Fibrous proteins

Fibrous proteins have a long, thin shape, and are often involved in structural roles.

Collagen structure

Collagen is a fibrous protein. It has a quaternary structure.

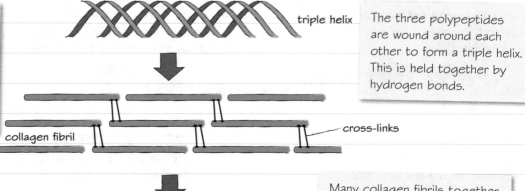

Cross-links form between adjacent triple helices, to form a collagen fibril. These cross-links are covalent bonds.

triple helix

The three polypeptides are wound around each other to form a triple helix. This is held together by hydrogen bonds.

collagen fibril

cross-links

Many collagen fibrils together form a collagen fibre.

collagen fibre

Fibrous protein function

Fibrous proteins are insoluble and have a structural role in the body.

Properties of collagen:

- mechanical strength
- part of bones, tendon, cartilage and connective tissue
- allows blood vessels to withstand high pressures.

Properties of keratin:

- many disulfide bridges
- hard and strong
- part of fingernails, hair, horns, hooves and feathers.

Properties of elastin:

- elastic
- part of skin, lungs, blood vessels and bladder
- allows structures to stretch and recoil.

Worked example

Compare the structures of haemoglobin and collagen.   **(5 marks)**

Both proteins have a quaternary structure.

Haemoglobin is made of four polypeptide chains and collagen has three.

Haemoglobin is a globular protein and collagen is a fibrous protein.

Haemoglobin has a prosthetic group called a haem group.

Collagen fibres are joined by cross-links to form collagen fibrils.

Remember cross-links are covalent bonds **not** hydrogen bonds.

Now try this

1 Why is collagen considered to be a quaternary protein?   **(2 marks)**
2 Suggest why collagen has great mechanical strength.   **(3 marks)**

Globular proteins

Globular proteins are roughly spherical, and usually have a role as an enzyme or hormone.

Haemoglobin structure

The role of haemoglobin is to carry oxygen in the blood. In the lungs, each haem group associates with one molecule of oxygen. In the body's tissues, these oxygen molecules disassociate with the haem group and move into the cells. Haemoglobin is a globular protein. It has a quaternary structure that is made up of four polypeptide chains.

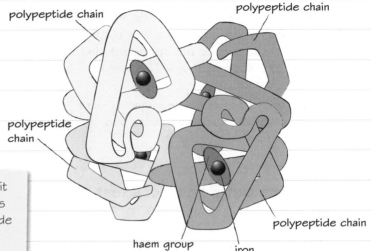

polypeptide chain

polypeptide chain

polypeptide chain

polypeptide chain

haem group

iron

Haemoglobin is a **conjugated** protein because it contains a **prosthetic group**. The haem group is an essential part of haemoglobin and is not made of amino acids; it contains an iron ion (Fe^{2+}).

Globular protein function

Globular proteins are soluble because of the polar amino acids on their outside surface. This means that they can be easily transported in the plasma, or move easily in the cytoplasm of a cell.

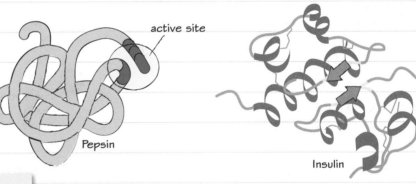

active site

Pepsin

Insulin

Enzymes, such as pepsin, and hormones, such as insulin, are globular.

How does the structure of haemoglobin aid its function?   **(3 marks)**

Contains haem groups which associate with oxygen.

Hydrophilic amino acids are on the outside of the haemoglobin molecule.

This makes haemoglobin soluble in the red blood cell/erythrocyte cytoplasm.

Each haem group can bind to one molecule of oxygen.

Hydrophilic amino acids attract water, which makes them soluble in water.

Remember an erythrocyte is a red blood cell.

1   What is a conjugated protein?   **(2 marks)**
2   Suggest why you might not expect to find disulfide bridges in haemoglobin.   **(2 marks)**

Benedict's test

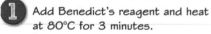 **Practical skills** Benedict's test is a test for sugars using Benedict's reagent. There are two different tests, depending on whether you have a **reducing** sugar or a **non-reducing** sugar.

Benedict's test for reducing sugars

A reducing sugar has an aldehyde group (CHO). Reducing sugars can reduce the copper(II) ions in the reagent into copper(I) ions. This forms a brick-red precipitate.

All monosaccharides are reducing sugars. Some common disaccharides, such as lactose and maltose, are also reducing sugars.

Benedict's test for non-reducing sugars

A non-reducing sugar does not contain a free aldehyde group. These sugars cannot reduce the copper(II) ions in the Benedict's reagent.

To test for a non-reducing sugar, you must first break the glycosidic bond. Benedict's reagent can now be used.

How to carry out Benedict's test

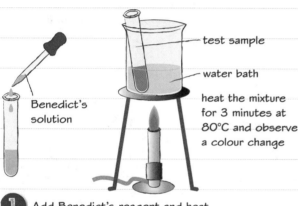

- test sample
- water bath

heat the mixture for 3 minutes at 80°C and observe a colour change

1 Add Benedict's reagent and heat at 80°C for 3 minutes.

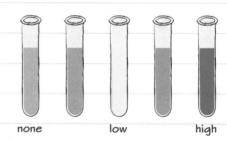

none   low   high

2 A brick-red precipitate indicates a high concentration of sugar. Green, yellow and orange indicate lower concentrations of sugar. A blue colour indicates no sugar present.

Non-reducing sugars

1 Add dilute HCl to the non-reducing sugar and boil for 5 minutes. Then neutralise with $NaHCO_3$.

2 Add Benedict's reagent and heat at 80°C for 3 minutes.

3 Green, yellow, orange and red indicate the presence of sugar.

Reagent test strips

Commercially produced test strips can be used to test for glucose. The strip is dipped into the solution and it changes colour. The colour can be compared to a chart that comes with the test and tells you the amount of glucose present. These are often used to test for glucose in the urine of diabetic patients.

Worked example

Which test would you use to test for the presence of sugar in milk? **(3 marks)**

Milk contains lactose, which is a reducing sugar. Therefore, I would use the Benedict's test for reducing sugars.

Now try this

1 When testing for sugars, what is the purpose of adding dilute HCl to sucrose? **(1 mark)**

2 What will happen to the sugar when the copper(II) ions are reduced? **(1 mark)**

Tests for protein, starch and lipids

 **Practical skills**   The presence of protein, starch or lipids can be detected with simple tests.

Biuret test for protein

The presence of protein can be tested using the biuret reagent. Copper ions in the biuret reagent react with nitrogen atoms in the peptide bond.

Add a few drops of the biuret reagent to the sample. If protein is present (+), the biuret reagent will change to lilac. A blue colour (−) indicates that there is no protein present.

Emulsion test for lipids

Lipids are not soluble in water, but they are soluble in alcohol.

Dissolving lipids in alcohol and then pouring the solution into water will make the lipids form an **emulsion** as they come out of solution in the water.

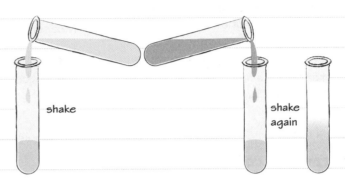

shake

shake again

Add ethanol to a sample of liquid and mix thoroughly. Tip the solution into a clean test tube containing water.

A white emulsion should form if lipids are present. The absence of any emulsion indicates that no lipids are present.

Iodine test for starch

Iodine solution, which contains potassium iodide, can be used to detect the presence of starch.

Add a few drops of iodine solution to the sample. If starch is present (+), the sample will change to blue-black. If there is no starch present (−), the iodine solution will stay yellow-brown.

Qualitative or quantitative?

All of these food tests are **qualitative** tests. This means that you only make an observation. You cannot measure the quantity of the protein, starch or lipid.

See page 20 for an example of a **quantitative** test.

See page 20 for an example of a **quantitative** test.

Worked example

Which food tests would cheese show a positive result for?   **(2 marks)**

Biuret test
Emulsion test   Cheese contains protein and lipid but no starch.

Now try this

1 Describe the observation you would expect when carrying out the emulsion test on milk.   **(2 marks)**
2 Explain how a food sample can be blue-black after the iodine test, but after incubation with an enzyme it shows a brick-red precipitate with the Benedict's test.   **(3 marks)**

Practical techniques – colorimetry

 Practical skills Colorimetry is a **quantitative** method. It can be used to work out the concentration of a solution.

What is a colorimeter?

A colorimeter is a machine that measures the absorbance of filtered light through a coloured solution. The higher the absorbance, the higher the concentration of the solution.

A light is shone through a coloured filter and then through the solution. A detector inside the colorimeter detects the amount of light that passes through the solution.

How does a colorimeter work?

The colorimeter can calculate the amount of light absorbed by the solution. This is shown on a digital display.

Measuring the absorbance of a range of solutions of known concentrations allows a calibration curve to be plotted. This graph can be used to read off the concentration of unknown solutions.

How to carry out colorimetry

Colorimeter with cuvette

1 A sample of distilled water is placed into the colorimeter to set it to 0.00. Samples of known concentration are added to the colorimeter.

2 The filter in the colorimeter is changed to the colour that would best be absorbed by the sample. The absorbance of each sample is measured.

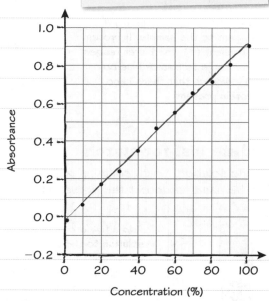

Colorimeter calibration curve

3 The absorbances of unknown samples are then measured and the concentration is worked out using the calibration curve.

Worked example

How is the concentration of a solution calculated using a colorimeter?   **(4 marks)**

The absorbances of a range of solutions of known concentration are measured using a colorimeter.
A calibration curve is drawn.
The absorbance of the solution of unknown concentration is measured using the colorimeter.
The concentration of the solution is read off the calibration curve.

 The known solutions are made of the same compound as the unknown solutions.

Biosensors

You can use a **biosensor** to measure the amount of a compound in solution. The biosensor converts the compound into an electrical signal that is proportional to the amount of compound.

Now try this

1 What does a colorimeter measure?   **(1 mark)**
2 What is the role of the detector in the colorimeter?   **(2 marks)**

Practical techniques – chromatography

Practical skills   Chromatography is a **qualitative** method. It is a technique that separates mixtures in order to detect their components.

Process of chromatography

Chromatography separates the components of a mixture using a **solvent**. The most common types are paper chromatography and thin layer chromatography. Both of these techniques work in a similar way, except thin layer chromatography uses a thin coating of powder on a glass plate.

1 A small amount of mixture is added to the bottom of the paper, which is just touching a solvent.

2 The solvent moves up the paper, carrying the components of the mixture with it.

3 Each component will bind to the paper with a different **affinity**.

- Components with the highest affinity will be at the bottom of the paper.
- Components with the lowest affinity will be at the top of the paper.
- Therefore, the components are separated according to their **differential affinities**.

Paper chromatography

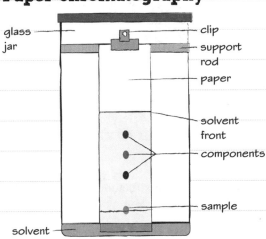

glass jar — clip, support rod, paper, solvent front, components, sample, solvent

The distance travelled up the paper by the solvent is called the **solvent front**. Each distance travelled by each component of the mixture is called the **solute front**.

Retention values

Each component of a mixture has its own retention (R_f) values which identify the compound. This can be worked out using the formula:

$$R_f = \frac{\text{Solute front}}{\text{Solvent front}}$$

For example, if the solute travels 10 cm and the solvent travels 15 cm, the R_f value is

$$\frac{10}{15} = 0.67$$

Applications of chromatography

- Pharmaceutical – test purity of products.
- Environmental science – to test for pollutants.
- Medicine – blood processing and purification.
- Forensic science – to identify compounds.

Many different solvents could be used, but the most common ones are water, ethanol or propanone (acetone).

Worked example

How would you use chromatography to work out the components of chlorophyll? **(4 marks)**

Add sample of chlorophyll to paper.

Place solvent beneath the paper.

Allow solvent to move up the paper and separate the components.

Use the solvent front and the solute fronts to work out the R_f values to identify the components.

Now try this

1 If a component moves 12.5 cm up the paper, and the solvent front moves 16 cm, what is the R_f value? **(1 mark)**

2 If the R_f value is 0.4 and the solvent front is 12 cm, how far did the component move? **(2 marks)**

3 Explain what is meant by differential affinity. **(3 marks)**

Exam skills

This question uses knowledge and skills that you have already revised. Look at pages 9, 10, 11 and 18 for a reminder about condensation reactions, glucose, glycogen and the Benedict's tests.

Worked example

Glucose is an important molecule in our bodies.

(a) Draw the structure of α-glucose.   **(2 marks)**

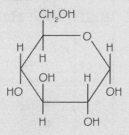

Remember to draw all of the carbon, oxygen and hydrogen atoms. If in doubt, count them up. There should be 6 carbon atoms, 6 oxygen atoms and 12 hydrogen atoms.

Make sure that you draw the hydroxyl groups and the hydrogen atoms in the correct configuration for each carbon. For carbons 1 to 4, the hydroxyl groups should all be below the carbon and the hydrogen atom above, except for carbon 3.

(b) How does this structure differ from β-glucose?   **(1 mark)**

β-glucose has the hydroxyl group (OH) and the hydrogen atom (H) the other way around on carbon 1.

(c) How does α-glucose form glycogen in the liver cells?   **(4 marks)**

Glycogen is made of many α-glucose molecules. One glucose molecule loses a hydroxyl group from carbon 1 and another glucose molecule loses a hydrogen atom from carbon 4, forming an α-1–4-glycosidic bond.

Remember the branches! Always include both types of glycosidic bond. This would also be true if answering a question about starch, as amylopectin also contains branches.

Many short branches of glucose molecules form at intervals along the glycogen chain. One glucose molecule loses a hydrogen atom from carbon 6 and another glucose molecule loses a hydroxyl group from carbon 1, forming an α-1–6-glycosidic bond.

(d) Glucose is a reducing sugar. Describe how an unknown solution could be tested for the presence of a reducing sugar.   **(5 marks)**

You should use the Benedict's test for reducing sugars. Add some Benedict's reagent to the unknown solution and heat for a few minutes. If the solution remains blue, there is no reducing sugar present. If the solution forms a green, yellow, orange or brick-red precipitate, then reducing sugars are present.

🧪 **Practical skills**   You will be asked about practical work in the exam. You need to remember the steps and reagents for all of the qualitative food tests.

A reagent test strip could also be used. If the strip changes colour, then reducing sugar is present.

Don't forget to include reagent test strips in your answer. If the question is looking for a quantitative answer, you should discuss colorimetry and biosensors.

Nucleotides

DNA and RNA are made of individual nucleotides. They share similar structures with some important differences.

DNA nucleotides

A DNA nucleotide is made from a nitrogenous base, a deoxyribose sugar and a phosphate group.

There are four different nitrogenous bases in DNA: adenine (A), cytosine (C), guanine (G) and thymine (T). Adenine and guanine are **purines** (a two-ring structure) and cytosine and thymine are **pyrimidines** (a one-ring structure). One purine always pairs with one pyrimidine.

RNA nucleotides

An RNA nucleotide is made from a nitrogenous base, a ribose sugar and a phosphate group.

The nitrogenous bases in RNA are the same as in DNA, except RNA contains uracil (U) in place of thymine.

The formation and breakage of phosphodiester bonds

The hydrogen (H) from the hydroxyl group on carbon 3 of the sugar of one nucleotide, and the hydroxyl group from the phosphate group of the other nucleotide are removed. A dinucleotide and a molecule of water are formed. This is a **condensation** reaction. A **phosphodiester bond** forms between carbon 3 of the sugar and the phosphate group. When many nucleotides are joined in this way, the resulting polymer is called a **polynucleotide**.

To break the phosphodiester bond, a molecule of water is added to the bond, which adds a hydrogen atom to the carbon 3 of the sugar and a hydroxyl group to the phosphate group. This is a **hydrolysis** reaction.

Compare and contrast the structures of RNA and DNA nucleotides.   **(3 marks)**

They both have phosphate groups, a nitrogenous base and a pentose sugar.

DNA nucleotides have a deoxyribose sugar and RNA nucleotides have a ribose sugar.

RNA nucleotides have uracil in place of thymine.

What is the sugar phosphate backbone of DNA made of?   **(2 marks)**

You need to know which carbons, on the pentose sugar, the phosphate group, the nitrogenous base and the phosphodiester bond are attached to.

Had a look ☐   Nearly there ☐   Nailed it! ☐

ADP and ATP

ATP is a phosphorylated nucleotide that is used in many reactions in the body. You need to know the structure of ATP and ADP molecules.

The structure of ATP

Adenosine triphosphate (ATP) has a similar structure to an RNA nucleotide. It has a nitrogenous base (adenine), a ribose sugar and three inorganic phosphate groups.

phosphate groups

adenine

ribose sugar

Uses of ATP

Area of the body	Function of ATP
Active transport across plasma membranes	ATP provides the energy for the carrier proteins to change shape and transport molecules against their concentration gradient (see page 38).
Muscles	ATP is attached to actin filaments in the muscle and provides the energy for the muscle to contract (see Module 5).
Glycolysis in cells	ATP provides the energy for pyruvate to be formed from triose phosphate (see Module 5).

How ATP is used as an intermediate energy store

The phosphoanhydride bond between the second and third phosphate is hydrolysed causing ATP to form ADP and an organic phosphate (Pi). This releases 30.5 kJ/mol of energy.

ATP is reformed from ADP and Pi during respiration or photosynthesis, where the third phosphate is added onto ADP.

Aerobic respiration makes 38 molecules of ATP per glucose molecule.

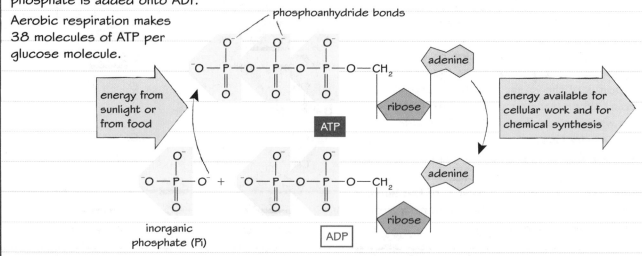

energy from sunlight or from food

phosphoanhydride bonds

ATP

adenine

ribose

energy available for cellular work and for chemical synthesis

inorganic phosphate (Pi)

ADP

adenine

ribose

Worked example

Compare and contrast the structures of ATP and a DNA nucleotide.   **(3 marks)**

Both contain a nitrogenous base, phosphate group and a pentose sugar. DNA nucleotide contains a deoxyribose sugar and ATP contains a ribose sugar. ATP has three phosphate groups.

Now try this

What type of reaction breaks the phosphoanhydride bond in ATP?   **(1 mark)**

The structure of DNA

Knowledge of the structure of DNA is very important for modern gene technologies. You should also be able to purify DNA by precipitation.

Deoxyribonucleic acid (DNA) is made up of many nucleotides joined together by phosphodiester bonds. This forms the sugar-phosphate backbone.

In between the two strands are the nitrogenous bases, adenine (A), cytosine (C), guanine (G) and thymine (T).

The bases pair up to form complementary **base pairs**. Adenine always pairs with thymine with two hydrogen bonds, and cytosine always pairs with guanine with three hydrogen bonds.

The two DNA strands are **antiparallel**, which means that the third and fifth carbons on the deoxyribose sugar are facing in opposite directions. As the two DNA strands twist, a **double helix** shape is formed.

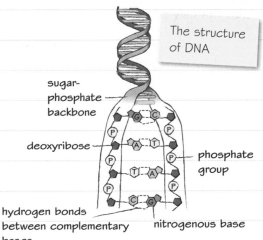

The structure of DNA

sugar-phosphate backbone

deoxyribose

phosphate group

hydrogen bonds between complementary bases

nitrogenous base

Practical skills   DNA is found in the nucleus of a cell. To extract and purify the DNA, the membranes need to be broken mechanically and chemically, using detergent and salt. The mixture is then filtered to remove cell debris.

DNA is often extracted from bananas and kiwi fruit in the classroom as they are **polyploid** (contain more than two sets of chromosomes).

Practical skills   **Precipitation of DNA**

The DNA can be precipitated from its solution by adding a layer of ethanol. The DNA will appear as a white precipitate.

Removing the salts in the solution can purify the DNA.

ethanol

| Tissue sample | Crush sample mechanically | Add detergent and salt | Filter mixture leaving DNA in solution | Add ethanol. DNA should form a white precipitate in the ethanol layer |

Worked example

Why does a purine always pair with a pyrimidine in the DNA helix?   **(2 marks)**

Maintains an equal distance between the strands of DNA.

Allows the DNA to twist into a double helix, increasing the stability of the molecule.

Discovery of DNA structure

James Watson and Francis Crick discovered the double helix structure of DNA in 1953, with significant contributions from Rosalind Franklin and Maurice Wilkins.

Now try this

1 Here is the DNA sequence from one strand of DNA:
   ATGTTTCTAACCAAGGGC
   What is the complementary DNA sequence?   **(2 marks)**
2 Why are the strands of DNA called the sugar-phosphate backbone?   **(2 marks)**

Semi-conservative DNA replication

How DNA replicates

Every time a cell divides, the DNA is replicated. This is done in a semi-conservative manner. That means that the new DNA always contains one strand of the original DNA, and one strand of new DNA.

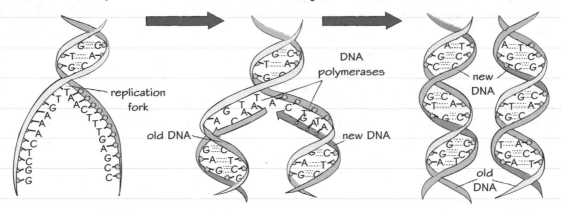

1 Enzymes called DNA helicases catalyse the breaking of the hydrogen bonds between the bases on the two original DNA strands.

2 DNA polymerases catalyse the addition of new DNA bases, complementary to the bases on each original strand.

3 Two DNA molecules are formed that are identical to the original DNA molecule.

Evidence for semi-conservative DNA replication

Early experimental evidence for semi-conservative DNA replication was carried out in *E.coli* bacteria.

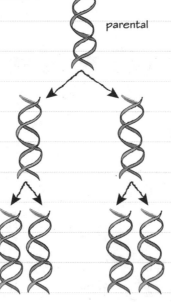

parental

after first replication cycle

after second replication cycle

Random spontaneous mutations

The DNA polymerases proofread the DNA bases to make sure that there are no mistakes. If a mismatch is found, the DNA polymerase removes the mismatched base and replaces it with the correct one. Mistakes that are unnoticed in the new DNA strands can lead to **mutations**.

Worked example

What are the roles of enzymes in DNA replication? **(3 marks)**

DNA helicase breaks the hydrogen bonds between the bases of the DNA strands.

DNA polymerase adds complementary bases to the free DNA strands.

DNA polymerase proofreads the bases to make sure that there are no mismatches.

Now try this

1 What is a mutation? **(1 mark)**
2 Explain what is meant by semi-conservative replication. **(2 marks)**

The genetic code

The genetic code is the sequence of bases in a gene. All organisms use the same four bases in their DNA – adenine, cytosine, guanine and thymine.

Features

In the genes, each set of three bases is called a **triplet**. Each triplet codes for a particular amino acid. The sequence of bases in a gene therefore codes for the sequence of amino acids in a protein. This sequence is the primary structure of the protein, as you covered on page 15.

Most amino acids have more than one triplet that codes for them. This is described as being **degenerate**.

The code is also non-overlapping, and so is read from a fixed point in groups of three bases.

The genetic code is **universal**. This means that the same triplets code for the same amino acids in all organisms. This is useful to geneticists because it means that genes can be taken out of one organism and placed in another.

Second base

	T	C	A	G	
T	TTT } Phe TTC TTA } Leu TTG	TCT TCC } Ser TCA TCG	TAT } Tyr TAC TAA Stop TAG Stop	TGT } Cys TGC TGA Stop TGG Trp	T C A G
C	CTT CTC } Leu CTA CTG	CCT CCC } Pro CCA CCG	CAT } His CAC CAA } Gln CAG	CGT CGC } Arg CGA CGG	T C A G
A	ATT ATC } Ile ATA ATG Met	ACT ACC } Thr ACA ACG	AAT } Asn AAC AAA } Lys AAG	AGT } Ser AGC AGA } Arg AGG	T C A G
G	GTT GTC } Val GTA GTG	GCT GCC } Ala GCA GCG	GAT } Asp GAC GAA } Glu GAG	GGT GGC } Gly GGA GGG	T C A G

First base ← (left)   Third base → (right)

> Using this table, the sequence of amino acids can be worked out from the sequence of triplets. The three-letter abbreviations are short for the name of the amino acid, for example Phe is short for phenylalanine.

Worked example

(a) What is the amino acid sequence coded for by the following DNA sequence?
ATGCTTTACATCAGACGTTGA   **(2 marks)**

Met-Leu-Tyr-Ile-Arg-Arg

(b) Why is there no amino acid for the last triplet?   **(1 mark)**

TGA is a stop triplet. It does not code for any amino acid.

> Stop triplets mean that it is the end of the gene.

(c) What would happen to the amino acid sequence if there were to be a base substitution in the second triplet, so that it read CTC instead of CTT?   **(1 mark)**

There would be no difference; the same amino acid would be coded for.

> This is called a **neutral**, or **silent, mutation**, as it makes no difference to the protein.

(d) What would happen to the amino acid sequence if there were to be a base deletion in the third triplet, so that it read TAA instead of TAC?   **(2 marks)**

The amino acid sequence would contain only two amino acids and would end at Leu, as TAA is a stop triplet.

Now try this

The genetic code is universal. Suggest how this could be useful to geneticists.   **(2 marks)**

Transcription and translation of genes

In order to make a protein, a gene has to be transcribed (copied) and then translated (read) by a ribosome. This involves an important enzyme, RNA polymerase, and several types of RNA.

Transcription

In order to transcribe a gene, the section of DNA containing that gene needs to be unwound, and the two strands separated.

RNA polymerase is an enzyme that adds free RNA nucleotides in the nucleus to exposed bases on one of the DNA strands (the **template strand**) in a complementary fashion. T (in DNA) bonds with A (in RNA), C bonds with G, and A (in DNA) bonds with uracil (U in RNA).

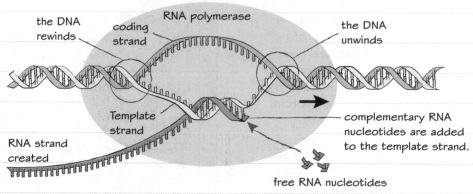

the DNA rewinds • coding strand • RNA polymerase • the DNA unwinds • Template strand • complementary RNA nucleotides are added to the template strand. • RNA strand created • free RNA nucleotides

This new RNA strand is called messenger RNA (mRNA). The mRNA is a copy of the information in the other DNA strand (the coding strand). The mRNA then moves out of the nucleus, through the nuclear pores, and attaches to a ribosome in the rough ER or in the cytoplasm.

Translation

The ribosome translates the mRNA. Ribosomes are made out of ribosomal RNA (rRNA). Each set of three bases on the mRNA is called a **codon**. Each codon codes for a particular amino acid, as you were reminded on page 27.

The mRNA strand fits into the space between the small and the large subunit of the ribosome, one codon at a time. The ribosome matches this codon to a sequence of three bases on a transfer RNA (tRNA) molecule. These three bases are called the **anti-codon**. On the other end of the tRNA is an amino acid. If the anti-codon is complementary to the codon in the ribosome, the amino acid is added on to the polypeptide chain. This process continues until the ribosome reaches the last codon of the mRNA strand, the **stop codon**.

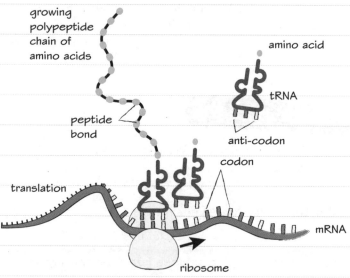

growing polypeptide chain of amino acids • amino acid • tRNA • peptide bond • anti-codon • codon • translation • mRNA • ribosome

Now try this

1 What are the similarities between the three types of RNA? **(2 marks)**

2 How are the structures of mRNA, rRNA and tRNA different? **(3 marks)**

The role of enzymes

Enzymes are proteins that catalyse reactions, both inside and outside cells.

What are enzymes

Enzymes are biological catalysts that speed up the rate of reactions, but are unused by the reaction.

Enzymes act within cells, and are also involved in making the structural parts of the human body, such as muscles, bone and connective tissue.

Enzymes work best at a specific temperature (37°C for human enzymes), normal pressure and a specific pH.

Intracellular enzymes

Some enzymes are intracellular. This means that they work inside the cell and catalyse metabolic reactions. One example of an intracellular enzyme is catalase.

Catalase breaks down harmful hydrogen peroxide in the liver, forming non-harmful products – oxygen and water.

$$\text{Hydrogen peroxide} \rightarrow \text{oxygen} + \text{water}$$
$$2H_2O_2 \qquad \rightarrow \qquad O_2 \quad + \quad 2H_2O$$

Activation energy

Enzymes work by lowering the amount of activation energy needed to carry out the reaction.

When the activation energy is lowered, the reaction can be carried out at a much lower temperature.

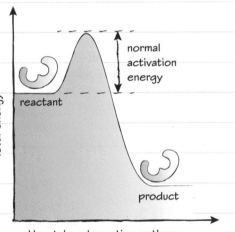

Uncatalyzed reaction pathway

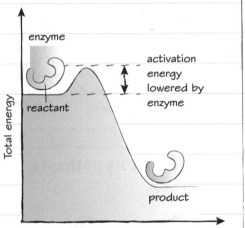

Enzyme-catalyzed reaction pathway

Extracellular enzymes

Extracellular enzymes work outside of the cells. They are made inside cells and then secreted (see page 38 for more about exocytosis). Many of our digestive enzymes are extracellular, for example, amylase and trypsin.

Amylase is found in saliva in the mouth, and breaks down starch into maltose. It is also found in the small intestine.

Trypsin is found in the small intestine and breaks down protein into amino acids.

Most enzymes are named for their substrate:
- lactase digests lactose
- maltase digests maltose
- cellulase digests cellulose.

'Intra' means inside and 'extra' means outside. 'Cellular' refers to cells.

Worked example

What is the difference between an intracellular and an extracellular enzyme? **(2 marks)**

Intracellular enzymes work inside cells.
Extracellular enzymes work outside of cells.

Remember catalase as an example of an intracellular enzyme and amylase and trypsin as examples of extracellular enzymes.

Now try this

1 Suggest the function of the following enzymes:
   (a) sucrase   (b) peptidase   (c) lipase.
   (3 marks)

2 Explain why enzymes are so useful in reactions.
   (3 marks)

The mechanism of enzyme action

Enzymes have a **specific** tertiary shape that allows them to bind to their substrate in a **complementary** way. There are two theories for how this happens: 'lock-and-key' and 'induced-fit'.

Enzyme action

Enzymes are globular proteins with a specific **active site**. Each active site will only fit a particular substrate. For example, catalase will only bind to hydrogen peroxide.

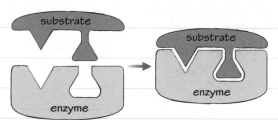

When a substrate is bound to the enzyme's active site, this is called an **enzyme–substrate complex**. The substrate becomes a product while still bound to the enzyme. This is called the **enzyme–product complex**. The enzyme is not used up by this reaction.

Lock-and-key hypothesis

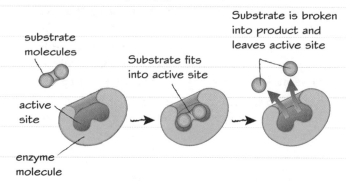

The active site of the enzyme is like a 'lock' and the substrate is like a 'key'. The two fit together in a complementary way. A particular substrate will only fit into a particular enzyme, in the same way that a key fits into a lock.

Induced-fit hypothesis

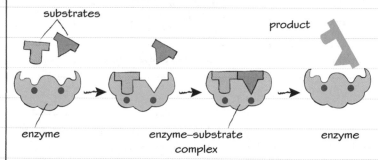

Research into enzyme structure before and after substrate binding, suggests that the shape of the enzyme is not rigidly fixed. The active site of the enzyme will only accept a particular substrate, and after the substrate binds the enzyme moulds around it to allow a tighter fit.

Worked example

How are enzymes specific to their substrate?

(2 marks)

The active site of the enzyme is a specific shape that is complementary to the shape of the substrate.

Only the substrate will fit into the active site.

 Always talk about **specificity**, the shape of the active site **and** the substrate, and the fact that these shapes are **complementary** to each other.

Remember to make this final point!

Now try this

1. Compare the induced-fit hypothesis to the lock-and-key hypothesis. **(2 marks)**
2. Why is the induced-fit hypothesis now the preferred hypothesis? **(2 marks)**

Factors that affect enzyme action

Enzymes act within a very specific environment. Any change in temperature, pH, substrate concentration or enzyme concentration can affect their reaction rate.

Effect of temperature

Most enzymes in the body have an **optimum** temperature of 40 °C. This means that they work at their fastest rate at this temperature.

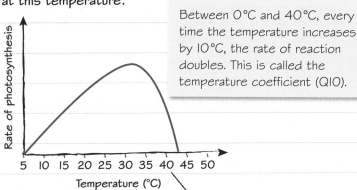

Between 0 °C and 40 °C, every time the temperature increases by 10 °C, the rate of reaction doubles. This is called the temperature coefficient (Q10).

Up to 40 °C: higher temperature, so more kinetic energy and more frequent collisions between enzyme and substrate. Rate of reaction increases.

Above 40 °C: the hydrogen bonds break in the **tertiary structure** of the active site. The enzyme **denatures**, and the substrate can no longer bind to the active site.

Effect of pH

Different enzymes in the body have a different optimum pH. For example, pepsin in the stomach has an optimum pH of 2, but trypsin in the small intestine has an optimum pH of 8.

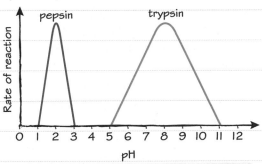

Above or below the optimum pH, hydrogen bonds holding together the tertiary structure of the active site are broken. The enzyme is denatured. The substrate is no longer able to bind to the active site.

Effect of enzyme concentration

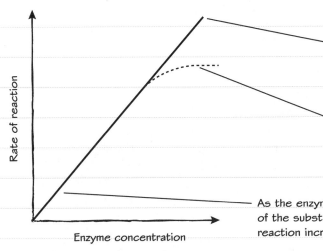

If the substrate is in excess, then the rate of reaction will continue to increase as the enzyme concentration increases.

If the substrate is limited, the rate of reaction will become constant when there is not enough substrate for the reaction to take place any faster.

As the enzyme concentration increases, there is more chance of the substrate colliding with the enzyme, and so the rate of reaction increases.

Worked example

What will happen to an enzyme if it is heated above 40 °C?
(3 marks)

The temperature will break the hydrogen bonds.

The tertiary structure of the active site will be denatured.

The enzyme will no longer be able to bind to its substrate.

Now try this

1 What happens to pepsin when the pH increases above pH 2?   **(3 marks)**

2 How does an increase in enzyme concentration affect an enzyme's rate of reaction in excess substrate?   **(2 marks)**

Factors that affect enzyme action – practical investigations

 Practical skills You will be expected to carry out at least one investigation into the effect of one factor on the rate of reaction of an enzyme.

The effect of substrate concentration on the rate of reaction

In this investigation, the enzyme catalase is catalysing the formation of water and oxygen from hydrogen peroxide (H_2O_2):

$$2H_2O_2 \mapsto 2H_2O + O_2$$

The volume of oxygen collected in 5 minutes is measured.

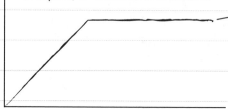

As the substrate concentration increases, there is more chance of catalase colliding with hydrogen peroxide to form an enzyme–substrate complex. Therefore, the rate of reaction increases. At this point some of catalase's active sites are not occupied, so catalase is in **excess**.

All the active sites of catalase are occupied; no further increase in rate is possible.

Rate of reaction (cm^3 O_2/min)

Hydrogen peroxide concentration (%)

Presenting results as a table

Units should always be in the column header (in parenthesis).

Make sure the independent variable is the the left column.

H_2O_2 concentration (%)	Rate of reaction (cm^3 O_2/min)			
	1	2	3	mean
0	0.0	0.0	0.0	0.0
2	2.0	1.5	2.5	2.0
4	3.5	4.5	4.0	4.0
6	6.2	6.0	5.8	6.0
8	6.0	6.1	5.9	6.0

Table should be in a box (drawn with a ruler).

All figures should be to the same number of decimal places.

To make an investigation valid, all of the variables should be kept constant except for the independent variable.

Independent variable – the variable that is changed, e.g. substrate concentration.

Dependent variable – variable that is being observed or measured, e.g. oxygen production.

Now try this

1 If one result is not concordant with the other results, it could be an anomaly. What should you do with an anomalous result?   **(2 marks)**

2 Suggest a limitation of this investigation and discuss what effect it would have on the data collected.   **(2 marks)**

Cofactors, coenzymes and prosthetic groups

Some enzymes need help from cofactors, coenzymes or prosthetic groups to carry out their function.

Cofactors

Cofactors are non-protein, inorganic substances that have to be present in order for the enzyme reaction to occur.

Cofactors can act as **co-substrates**. They bind to an inactive protein and allow the substrate to bind to the active site.

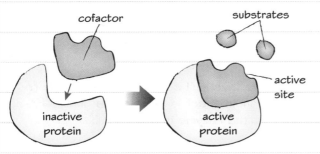

Cofactors can also alter the **charge** on the surface of the substrate or active site of the enzyme.

Cofactors are usually **mineral ions**. They do not permanently bind to the enzyme, but they temporarily bind to the enzyme–substrate complex, to make its formation easier.

Coenzymes

Coenzymes are a type of small, organic cofactor. They are used up by the enzyme reaction, and have to be recycled. Coenzymes help the enzyme by binding temporarily to the active site with the substrate and carrying chemical groups between enzymes.

coenzyme (vitamin or mineral)

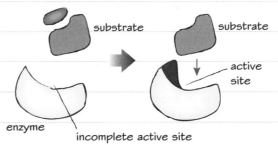

Many coenzymes are derived from vitamins. This is why it is so important to include them in our diet.

Apoenzymes and holoenzymes

An enzyme without its cofactor is called an apoenzyme. When the apoenzyme is bound to its cofactor, it is known as a holoenzyme.

Worked example

Compare coenzymes and cofactors.   **(3 marks)**

Cofactors are inorganic substances, e.g. mineral ions, whereas coenzymes are organic molecules, e.g. vitamins.

Cofactors can be a permanent part of the enzyme, e.g. a prosthetic group, whereas coenzymes bind to the active site of an enzyme temporarily.

Cofactors speed up the enzyme's rate of reaction, whereas coenzymes are intermediate carriers between enzymes.

Cofactors and prosthetic groups are non-protein mineral ions. Coenzymes are non-protein organic molecules.

Co-substrates also bind to the active site temporarily, but they are a mineral ion and so **not** a coenzyme.

Prosthetic groups

Prosthetic groups are cofactors that are a permanent part of the enzyme. Some examples of prosthetic groups include:

- haem group (includes Fe^{2+}) in haemoglobin
- magnesium ion (Mg^{2+}) in chlorophyll
- zinc ion (Zn^{2+}) in carbonic anhydrase.

zinc prosthetic group

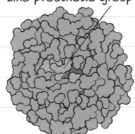

carbonic anhydrase

Now try this

Suggest how mineral ions altering the charge on the surface of the active site allows the enzyme–substrate complex to form more easily.   **(2 marks)**

Inhibitors

Inhibitors bind to the enzyme and slow down or stop the enzyme's activity.

Competitive inhibitors

Competitive inhibitors bind to the active site of the enzyme. They compete with the enzyme for access to the active site. Once the inhibitor is bound, the substrate cannot bind.

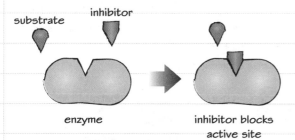

Most competitive inhibition is **reversible**. When the substrate is in excess, the substrate out-competes the inhibitor and rate of reaction returns to normal.

Non-competitive inhibitors

Non-competitive inhibitors bind to the enzyme's **allosteric** site. When the inhibitor binds, it alters the tertiary structure of the enzyme and distorts the active site so that the substrate can no longer bind.

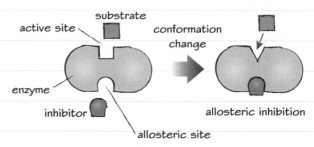

Many non-competitive inhibitors bind permanently to the enzyme so their effect is **irreversible**. In this case, we can say that the enzyme is denatured.

Inhibitors as poisons

Some poisons work by binding to enzymes. For example, **cyanide** is a non-competitive inhibitor of the enzyme cytochrome oxidase. This enzyme is essential in aerobic respiration.

Cyanide poisoning is reversible, but the antidote must be given quickly.

Inhibitors as medicines

Many medicinal drugs, such as blood pressure medicines, work by inhibiting enzymes.

Angiotensin converting enzyme (ACE) inhibitors inhibit ACE. This prevents ACE from taking part in a **metabolic pathway** that would increase blood pressure.

Worked example

Compare the action of competitive and non-competitive inhibitors.　　**(3 marks)**

Competitive inhibitors bind to the active site of enzymes and block the substrate from binding.

Non-competitive inhibitors bind to the allosteric site of the enzyme. This changes the tertiary structure of the enzyme and distorts the active site, preventing the substrate from binding.

Both forms of inhibition prevent the substrate from binding.

Remember that although competitive inhibitors tend to be reversible and non-competitive inhibitors tend to be irreversible, this is not always the case.

Pepsinogen to pepsin

Some enzymes involved in metabolism are produced as inactive precursors that are activated when certain conditions are met. For example, inactive pepsinogen is activated to form pepsin when it is secreted into the acid environment of the stomach.

Now try this

1　Give an example of a poison that is an inhibitor, and its mechanism of action.　　**(3 marks)**

2　Explain how an inhibitor could be used as a medicinal drug.　　**(3 marks)**

Exam skills

This question is about enzymes and factors that affect their activity. Look at pages 29–34 to remind yourself about enzymes, their mechanism of action and the effect of inhibitors.

Worked example

Enzymes are biological catalysts. They catalyse many reactions both inside and outside of cells.

(a) What is meant by the term 'biological catalyst'?

(1 mark)

A biological catalyst is a molecule that speeds up a metabolic process in a living organism, without getting used up.

> Remember, a catalyst is any molecule that speeds up a reaction without being used up.

(b) Name an enzyme that works extracellularly.   **(1 mark)**

amylase / trypsin / pepsin

> You have to remember the examples of amylase and trypsin as extracellular enzymes.

(c) How is an enzyme specific for its substrate?   **(2 marks)**

The **active site** of the enzyme has a particular shape that is **complementary** to the shape of the substrate. The substrate fits into the active site of the enzyme.

> You should recall the complementary shapes of the enzyme's active site and a substrate as being like a lock and a key.

(d) The effect of temperature on enzyme activity is being investigated using amylase. Draw the expected shape of the graph.   **(2 marks)**

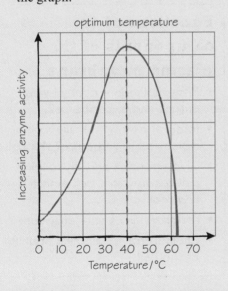

> Between 0°C and 40°C, the shape of the graph should be a steady increase as the rate of reaction doubles every 10°C.
>
> Between 40°C and 60°C, the shape of the graph should be a steep decline, as the rate of reaction eventually reaches zero.
>
> If you are asked to **describe** the shape of the graph, remember to describe the shape before **and** after 40°C.

(e) A competitive inhibitor of amylase is added to the reaction mixture. Explain how competitive inhibitors decrease the rate of reaction of the enzyme.   **(3 marks)**

The shape of the competitive inhibitor is complementary to the shape of the active site of the enzyme, and competes with the substrate to enter the active site. Once the competitive inhibitor binds to the active site, the substrate cannot bind there and form an enzyme–substrate complex. Therefore, no products can be made, and the rate of reaction decreases.

> Remember, the competitive inhibitor has a **similar** shape to the substrate, not the **same** shape.
>
> Competitive inhibition is usually reversible. Adding more substrate can increase the rate of reaction, as the substrate outcompetes the inhibitor.

The fluid mosaic model

The plasma membrane is made up of many different molecules. Each molecule has a different role.

The fluid mosaic model is a model of the plasma membrane. The main components of the plasma membrane are **phospholipids**. Phospholipids have a **hydrophilic** head that is attracted to water, and a **hydrophobic** tail that repels water. Phospholipids form a **bilayer**.

The bilayer contains many proteins. Some are **integral** and cross the membrane, and some are **peripheral** and are found only on one side of the membrane.

Some components of the plasma membrane are attached to long carbohydrate chains. Phospholipids attached to chains are called **glycolipids**. Proteins attached to chains are called **glycoproteins**. The plasma membrane also contains **cholesterol**.

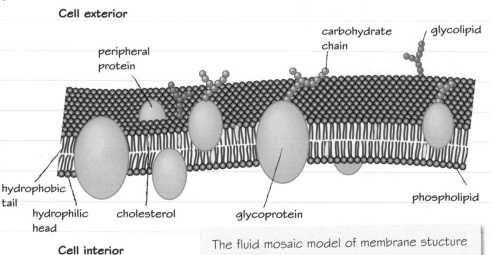

The fluid mosaic model of membrane stucture

Role of the plasma membrane

The plasma membrane around the cell:

- is partially permeable, and so controls the transport of molecules in and out of the cells
- separates the contents of the cell from the outside environment
- allows other cells to recognise the cell as belonging to the body
- allows molecules, such as hormones or drugs, to bind to the receptors in the membrane
- holds the components of some chemical reactions, e.g. enzymes, in place.

There are also plasma membranes around the organelles, which separate the contents of the organelles from the cytoplasm and allow molecules to be transported in and out of the organelle.

Functions of the components of the plasma membrane

- Phospholipids – control what enters and exits the cell.
- Integral protein – transport of molecules in and out of the cell.
- Glycoproteins and glycolipids – cell recognition, cell signalling and receptors.
- Peripheral protein – enzymes.
- Cholesterol – maintains stability of the cell.

Worked example

Why do the phospholipids form a bilayer? **(3 marks)**

Phospholipids have a hydrophilic head and a hydrophobic tail.

The hydrophobic tails repel water so form the inside of the bilayer.

The hydrophilic heads attract water and form the outside of the bilayer.

Now try this

Explain the role of plasma membranes around the mitochondria.

(2 marks)

Factors that affect membrane structure

Different plasma membranes contain different amounts of cholesterol, which affects the fluidity of the membrane. Other factors that affect the membrane structure and permeability are temperature and solvents.

The effect of temperature

As the temperature increases, the components of the plasma membrane have more **kinetic energy**. This means that phospholipids and proteins move about faster, and move further apart. This results in plasma membranes increasing their **fluidity** and **permeability**.

The proteins may move too far apart to carry out a chemical reaction. If the temperature increases above 40 °C, the proteins will become denatured.

As the temperature decreases, the kinetic energy of the components of the plasma membrane decreases, and the components move less. This maintains the fluidity of the membrane.

Betalain

Beetroot contains a red pigment called betalain. When sections of beetroot are placed in water of increasing temperatures the membrane becomes more permeable and the betalain leaks out of the cells and into the water.

Cholesterol in membranes

Cholesterol is embedded between the phospholipids in the plasma membrane and maintains the **mechanical stability** of the membrane. It does this by partially immobilising nearby phospholipids, and decreasing the permeability of the membrane to small water-soluble molecules.

Organisms that live at very cold temperatures have more cholesterol in their membranes. This is because the cholesterol molecules prevent the phospholipids from packing too closely together and freezing.

The effect of solvents on membranes

Some solvents, such as **acetone** and **ethanol**, dissolve lipids and so will permanently damage plasma membranes.

Thermophilic bacteria

Some organisms, such as thermophilic bacteria, can withstand extremely high temperatures. These organisms contain heat-tolerant lipids in their membranes, which make the membranes less permeable at higher temperatures.

Worked example

Why does the red pigment leak out of beetroot cells at high temperatures? **(3 marks)**

Increased kinetic energy.

Increased movement of phospholipids / phospholipids move further apart.

Membrane becomes more fluid / permeable.

Proteins would also denature at high temperatures, but this is unlikely to have an effect on the permeability of the plasma membrane.

Now try this

1 How does cholesterol prevent membranes from freezing? **(2 marks)**
2 What would be the effect of adding ethanol to beetroot membranes? **(3 marks)**

Movement across the membrane

The plasma membrane is **partially permeable** and molecules can cross it in a number of ways.

Diffusion

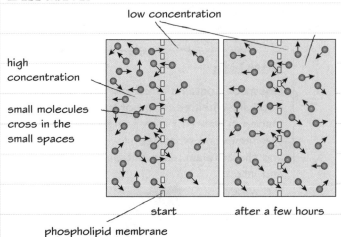

low concentration

high concentration

small molecules cross in the small spaces

start   after a few hours

phospholipid membrane (partially permeable)

Facilitated diffusion

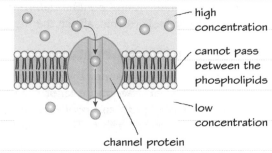

high concentration

cannot pass between the phospholipids

low concentration

channel protein

Larger or polar molecules, e.g. glucose, can cross the plasma membrane by **facilitated diffusion**. This is also a passive process.

Active transport

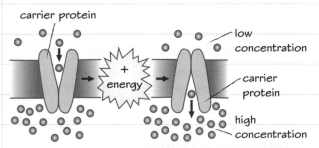

carrier protein

low concentration

+ energy

carrier protein

high concentration

When a cell needs to take in molecules against the **concentration gradient** this requires ATP as an immediate source of energy. This is an **active** process. An example of this happening is plants taking in potassium ions from the soil.

Bulk transport

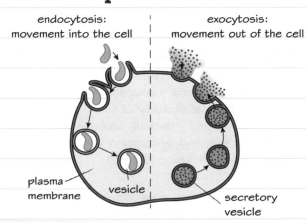

endocytosis: movement into the cell

exocytosis: movement out of the cell

plasma membrane   vesicle

secretory vesicle

Large molecules, such as proteins, have to be taken in and out of the cells by **bulk transport**, an active process.

The engulfing of pathogens by phagocytes is a form of endocytosis called **phagocytosis**.

Worked example

Polar means that the molecule carries a positive or negative charge.

Why must larger or polar molecules cross the plasma membrane by facilitated diffusion?   **(2 marks)**

Larger molecules are too big to fit between the phospholipids. The lipid tails of the phospholipids repel polar molecules.

Small, non-polar molecules, e.g. oxygen, and lipid molecules can cross the plasma membrane by **diffusion**. This process does not require any ATP so it is a **passive** process.

Now try this

1 Explain why lipid molecules can cross the plasma membrane by diffusion.   **(2 marks)**
2 What would happen to active transport in cells in a low-oxygen environment?   **(3 marks)**

Osmosis

Osmosis is the movement of water across the plasma membrane.

Movement of water

Water is a small, dipolar molecule. It can cross the plasma membrane by a special type of diffusion called **osmosis**.

Osmosis is the movement of water from an area of high water potential to an area of low water potential. The symbol for water potential is y.

Water potential is a measure of how free the water molecules are to move. Water potential is a pressure and is measured in kilopascals (kPa).

Pure water contains no solutes so has a high water potential. The highest possible water potential is 0 kPa.

A concentrated solution has many solutes, so the water potential is lower. It can also be described as being more negative, and is given a negative number e.g. −100 kPa.

Osmosis across a partially permeable membrane

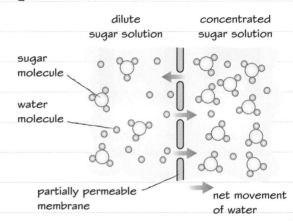

dilute sugar solution   concentrated sugar solution

sugar molecule

water molecule

partially permeable membrane

net movement of water

The dilute sugar solution has a higher water potential than the concentrated solution. The net movement of water is from the dilute solution to the concentrated solution.

The effect of osmosis on plant and animal cells

When cells are put into a solution with a higher water potential than inside the cell (hypertonic), water moves into the cell. Animal cells will burst. Plant cells have a strong cellulose cell wall – the increased volume of water makes the cells turgid.

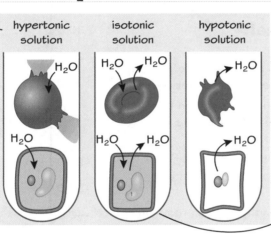

hypertonic solution   isotonic solution   hypotonic solution

H_2O H_2O H_2O H_2O

H_2O H_2O H_2O H_2O

When cells are put into a solution with a lower water potential than inside the cell (hypotonic), water moves out of the cell. Animal cells will shrink or **crenate**. The plasma membrane of plant cells comes away from the cell wall (**plasmolysis**).

When cells are put into a solution that has the same water potential as inside the cell (isotonic), water moves equally in both directions.

Worked example

A plant cell has a water potential of −100 kPa and is placed into a solution of −500 kPa. What will happen to the plant cell and why?   **(3 marks)**

The solution has a lower / more negative water potential than inside the cell.

Water will have a net movement out of the cell, from high to low water potential.

The plant cell will be plasmolysed.

Now try this

What would happen to a red blood cell placed into a hypotonic solution?   **(3 marks)**

Water potential can be described as higher or lower, or more or less negative. The closer to 0 kPa, the higher or less negative the water potential. The further way from 0 kPa, the lower, or more negative the water potential.

Factors affecting diffusion

You will be expected to do at least one investigation into factors that affect diffusion rates in a model cell.

 Maths skills **The effect of surface area on diffusion**

Agar cubes of different sizes made up with hydrochloric acid (HCl) and phenolphthalein indicator.

1 Cubes are cut from agar/HCl, with sides of 2 cm, 1 cm and 0.5 cm. Place them into a beaker of sodium hydroxide (NaOH).

2 As the NaOH diffuses into the agar cube, the colour of the cube should change from clear to bright pink due to the presence of the phenolphthalein indicator.

3 Remove cubes after 5 minutes and cut them open. Measure the width of the pink zone.

4 Work out the rate of diffusion:

$$\text{rate of diffusion (cm/min)} = \frac{\text{width of pink agar (cm)}}{\text{time (min)}}$$

Practical skills **Determining the concentration of sugar in a cell**

1 Potato chips of the same mass and shape are placed into solutions of different sucrose concentration.

2 If the water potential is higher in the solution than the potato chip, then water will move into the potato cells and cause them to swell.

3 If the water potential is lower, then the water will move out of the potato cells and cause them to shrivel.

4 By plotting the results on a graph, you can read off the point at which the solution is isotonic with the water in the potato cells. This is the sucrose concentration of the cells.

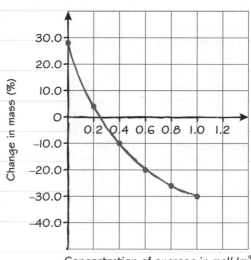

Graph of potato mass change and sugar concentration.

The NaOH takes the following amount of time to reach the centre of the three cubes (in size order from smallest) – 5 minutes, 10 minutes, 40 minutes. What conclusions can you make about diffusion in these cubes?   **(3 marks)**

All the cubes have the same rate of diffusion, 0.05 cm/min. The larger the surface area, the shorter the time for the NaOH to diffuse to the centre.

Now try this

What is the sucrose concentration of the potato cells?   **(2 marks)**

The cell cycle

Cells go through the cell cycle many times in their lives. At the end of each cycle, the cell divides by **mitosis** (see page 42), producing genetically identical cells.

There are four phases to the cell cycle: G_1, S, G_2 and M.

- G_1 and G_2 are growth phases and the S phase is when DNA is synthesised. Altogether, these three phases are called **Interphase**. Cells spend most of the cell cycle in interphase.

- The M phase is mitosis.

Not all cells can go through mitosis. Some stop at the checkpoints because there is something wrong in the cell. Other times, the cell rests in the G_1 phase.

Most cells go through the cell cycle a finite amount of times. The exact number depends on the cell type and is called the Hayflick limit. Only stem cells and tumour cells have no Hayflick limit. We say that these cells are **immortal** because they can go on dividing indefinitely.

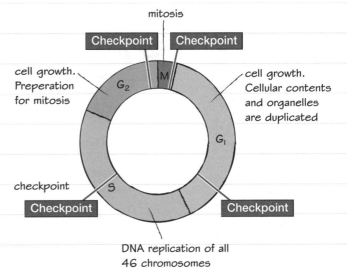

How the cell cycle is regulated

The cell cycle has several **checkpoints** where the DNA of the cell is checked. If the DNA is undamaged, the cell will be allowed to go through to the next phase.

The two main checkpoints are the G_1/S checkpoint and the G_2/M checkpoint. Here, a regulatory protein called p53 checks the DNA for damage, such as mismatched bases or mutations.

p53 is a **tumour suppressor gene**. It prevents mutated cells from dividing and becoming cancerous. If p53 itself is mutated, then cells will pass through these checkpoints without being checked and can quickly become cancerous.

Worked example

What are the roles of interphase?   **(3 marks)**

Growth of the cell
Duplication of organelles
DNA replication

G_0 phase

Sometimes cells enter a phase from the G_1 phase, called G_0. During this phase, cells can undergo differentiation or senescence (cells no longer divide). Cells can also undergo programmed cell death (apoptosis).

Now try this

1 Why are stem cells and tumour cells described as being immortal?   **(2 marks)**
2 How do checkpoint proteins prevent cells becoming cancerous?   **(2 marks)**

Mitosis

Mitosis is the division of **diploid** cells (cells with two sets of chromosomes) to give two daughter cells that are **genetically identical**.

The four stages of mitosis

prophase

metaphase

anaphase

telophase

During prophase, the duplicated chromosomes **supercoil** and become visible and the **nuclear envelope** breaks down. The **centrioles** start making the spindle fibres.

In metaphase, the chromosomes line up in the centre of the cell (the equator) and the centromere of the duplicated chromosome is attached to two spindle fibres.

During anaphase, the spindle fibres shorten, pulling the two halves of the duplicated chromosomes apart. These halves are called **sister chromatids**.

In telophase, each half of the cell now contains a full set of sister chromatids (now called chromosomes) and a new nuclear envelope forms around them. The cell divides to make two genetically identical daughter cells in a process called **cytokinesis**. Each new cell has a full set of chromosomes.

Practical skills Mitosis in plant cells

All of the stages of mitosis can be seen by observing prepared slides of dividing plant tissue. A good tissue in which to see this is the root tips of onions.

You may be asked to observe mitosis in plant tissue in the classroom.

The significance of mitosis in life cycles

Mitosis is needed in the body for growth, tissue repair and cell replacement.

* Children grow by increasing the number of cells in their body.
* When you injure your body, cells divide by mitosis to replace the damaged tissue.
* Some cells, e.g. skin cells, are regularly replaced by mitosis.

Some organisms divide by mitosis. This is called **asexual reproduction**. Plants, yeasts and fungi can all reproduce in this way.

The resulting offspring are genetically identical and can be described as **clones**.

Worked example

From looking at a slide of dividing plant tissue, how can you tell that a cell is in anaphase? **(2 marks)**

The sister chromatids are separated.
The chromatids are moving towards opposite ends of the cell.

Now try this

Describe what happens to the chromosomes in each stage of mitosis. **(4 marks)**

Each half of the duplicated chromosome is called a sister chromatid. When the sister chromatids separate, they are referred to as chromosomes again.

Meiosis

Meiosis is a type of cell division that makes **haploid** cells (cells with half the full number of chromosomes). These are used as **gametes** (sex cells). Meiosis involves two rounds of division.

Meiosis 1

prophase I   metaphase I   anaphase I   telophase I

- Homologous chromosomes become visible and pair up on the equator of the cell.
- Each chromosome is attached to spindle fibres.
- During metaphase I, each pair of homologous chromosomes is separated.
- The first round ends with two haploid daughter cells that are not genetically identical.

Meiosis 2

prophase 2   metaphase 2   anaphase 2   telophase 2

The second round is similar to mitosis.

- The chromosomes line up on the equator in 'single file' and each sister chromatid is separated by the spindle fibres.
- At the end of the second round, there are four haploid daughter cells that are not genetically identical.

Homologous chromosomes

You have two copies of each chromosome in every cell; one from your mother and one from your father.

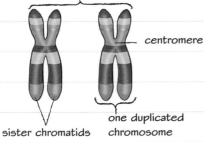

homologous pair of chromosomes

centromere

sister chromatids   one duplicated chromosome

Although each homologous chromosome contains the same genes, in the same order, they will not necessarily have the same **alleles** (version of a gene). For example, both of the homologous chromosomes may have the gene for blood type, but one may have the blood type A allele and the other may have the blood type O allele.

The significance of meiosis in life cycles

Meiosis increases genetic variation in two important ways:

1. Independent assortment – homologous chromosomes can be randomly assorted to either end of the cell when separated during metaphase I.

2. Crossing over – homologous chromosomes cross over strands while in prophase I and exchange genetic material. The point at which they cross is called a **chiasma**.

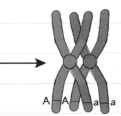

Crossing over can occur between these chromosome strands

chiasma

chiasma

A a A a   A A a a

The genes on each homologous chromosome are still the same but the alleles have been exchanged.

Worked example

Contrast meiosis and mitosis.   **(3 marks)**

Meiosis has two rounds of division and mitosis has one round of division. Mitosis produces two genetically identical daughter cells and meiosis produces four non-identical daughter cells. The daughter cells in mitosis are diploid and the daughter cells in meiosis are haploid.

Think of a mnemonic to remember the order of the stages of meiosis, for example Pandas Meander Around Trees.

Now try this

How does meiosis increase genetic variation?   **(2 marks)**

Specialised cells

Cells in multicellular organisms are specialised to perform a particular function.

Specialised animal cells

Red blood cells (**erythrocytes**) carry oxygen around the body. They are biconcave, have no nucleus and are packed with haemoglobin.

Neutrophils are phagocytes. Their nuclei are lobed so that they can squeeze through gaps in the capillary wall, into the tissues. Their cytoplasm is packed with lysosomes to digest any pathogens.

Sperm cells are gametes that swim to the ovum (egg) using an undulipodium (a type of flagellum). The head of the sperm contains enzymes to burrow through the outer layer of the ovum.

Specialised plant cells

Palisade cells are found near the top surface of the leaf. They are long and thin so that many cells can be packed close together. They are filled with chloroplasts to absorb sunlight for **photosynthesis**.

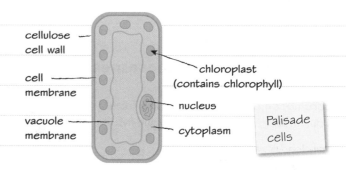

cellulose cell wall — chloroplast (contains chlorophyll)

cell membrane

nucleus

vacuole membrane — cytoplasm

Palisade cells

Guard cells are found on the underside of the leaf. They come in pairs, and control the opening and closing of the **stomata**.

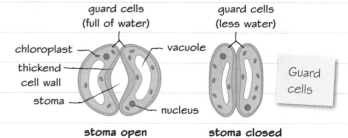

guard cells (full of water)   guard cells (less water)

chloroplast — vacuole

thickend cell wall

stoma — nucleus

stoma open   stoma closed

Guard cells

Root hair cells are found in the roots. They have long thin hairs to increase their **surface area** for taking up water and minerals from the soil.

Root hair cells

How do guard cells control the opening of the stomata? **(3 marks)**

Water enters the guard cells by osmosis, which makes the guard cells swell.

The cellulose cell wall of the guard cells is thicker and more rigid along the body of the cell.

The cells swell more at the tips than the body of the cell, causing the stoma to enlarge.

The guard cells will only open the stomata when there is sufficient water. The stomata close when there is less water because water no longer enters the guard cells. Closing the stomata helps the plant to avoid water loss.

Compare the structures of erythrocytes and neutrophils. **(3 marks)**

Specialised tissues

Tissues in multicellular organisms are specialised to perform a particular function. All of the cells in a tissue work together to perform a similar function.

Specialised animal tissues

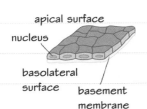
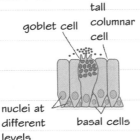
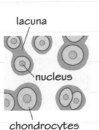
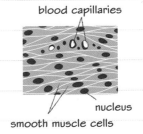

Squamous epithelium is a tissue made of smooth, flattened cells. It lines the inside of blood vessels and alveoli.

Ciliated epithelium is a tissue made of ciliated cells and **goblet cells**. The goblet cells produce mucus and the **cilia** waft the mucus away from the lungs.

Cartilage is a strong tissue that is used to make tendons, bones and connective tissue.

Muscle tissue is made of many muscle cells that all contract in the same direction.

Specialised plant tissues

Xylem tissue transports water and minerals from the roots to the top of the plant. It is made of many xylem elements joined end to end to form a long, hollow tube. Xylem tissue is **dead**. The cell walls are lined with a strong, waterproof material called **lignin**.

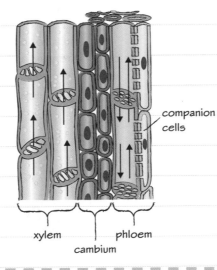

Phloem tissue transports sugars (mainly **sucrose**) and other assimilates up and down the plant. It is made of many sieve elements joined end to end, separated by **sieve plates**. Phloem is living tissue, but supported by **companion cells**.

Worked example

What is a tissue?   **(2 marks)**

A group of similar cells that work together to perform a particular function.

From cells to organs

Cells are organised into tissues. Tissues are organised into organs. Organs are organised into organ systems.

Organs are groups of tissues working together to perform a particular function.

Now try this

How is xylem specialised for its function?   **(3 marks)**

Stem cells

Stem cells are undifferentiated cells that can differentiate into any cell type.

What are stem cells?

Stem cells are found in embryos and, to a lesser extent, in adults.

Cells in the very early embryo are **totipotent** stem cells. This means that they can differentiate into any cell type in the body or the placenta.

The embryo then develops into a blastocyst.

The inner cell mass is made of **pluripotent** stem cells. These cells can differentiate into any cell type in the body.

Adult stem cells are fewer in number and are **multipotent**. This means that they can only differentiate into a small number of cell types.

Embryonic stem cells

Treatment with different chemicals stimulates the stem cells to differentiate into different cell types.

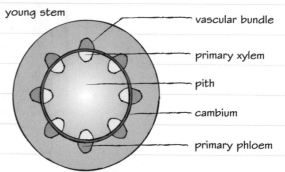

inner cell mass — trophoblast

fluid filled blastocyst cavity

Cells from the inner mass of the blastocyst are used in stem cell research.

Animal stem cells

Erythrocytes (red blood cells) and neutrophils are made from adult stem cells in the bone marrow in the long bones.

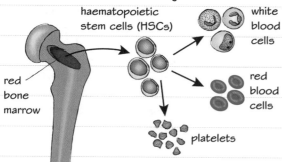

haematopoietic stem cells (HSCs)

white blood cells

red blood cells

red bone marrow

platelets

The haematopoietic stem cells in the bone marrow can differentiate into any blood cell.

Plant stem cells

Plant stem cells are found in the **meristem** tissue in plants. This is found at the root and shoot tips and in the **cambium** (an area of multipotent stem cells between the xylem and the phloem). Xylem and phloem are formed from the cambium.

young stem

vascular bundle

primary xylem

pith

cambium

primary phloem

Young phloem and xylem are packed together either side of the cambium tissue in a **vascular bundle**.

Worked example

What is the difference between multipotent and pluripotent?   **(2 marks)**

Multipotent cells can only differentiate into a small number of cell types. Pluripotent cells can differentiate into any type of body cell.

Adult stem cells are multipotent. The inner mass cells of an embryo are pluripotent.

Now try this

1   What is a stem cell?   **(2 marks)**
2   Explain how neutrophils are made in the bone marrow.   **(3 marks)**

Uses of stem cells

Stem cells have many potential uses in research and medicine.

Uses of stem cells

Embryonic stem cells are often used in research because they are pluripotent (look back at page 46 for a definition of pluripotent), and plentiful. However, using these cells raises ethical concerns, as these embryos could have gone on to develop into babies. For this reason stem cell research is banned in some countries.

Adult stem cells can be used for research, but these cells are multipotent and are fewer in number.

Stem cells have been used in several areas of medicine to improve the quality of life and even to save lives. This raises another ethical concern. Is it acceptable to let people suffer or die when you could have treated them using stem cells?

As scientific techniques improve, fewer embryonic stem cells may be needed for research in the future.

Repair of damaged tissues

Stem cells have been successfully used to grow new organs, or parts of organs for transplant. The patient's adult stem cells are removed and stimulated to differentiate into a particular cell type. These cells are placed over a scaffold, and grow into the shape of the organ they are to replace. So far, this has been done with simple organs and tissues.

Neurological conditions

Research is currently being done on whether stem cells could be used to make new neurones in the brain, reversing the effects of some neurological disorders, such as Alzheimer or Parkinson's disease.

It is hoped that by replacing the neurones, the memory loss and dementia will be reversed. There has been some success in rats using this technique.

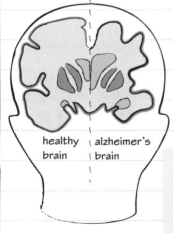

healthy brain | alzheimer's brain

Alzheimer disease destroys neurones in the brain, leading to memory loss and dementia.

Developmental biology

Embryonic stem cells can be used to study the development of the embryo. This information is important because it tells us what chemical signals are needed in order to make a stem cell differentiate into a specialised cell.

Induced pluripotent stem cells (iPS) have been made using the information from this research. Differentiated cells have been treated with chemicals to make them undifferentiate back into adult stem cells.

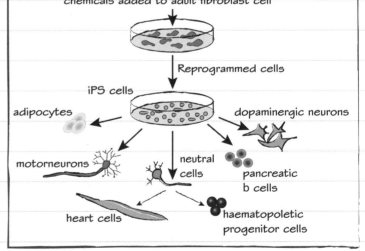

Worked example

Where can stem cells be found, besides in developing embryos? (3 marks)

adult stem cells   amniotic fluid   umbilical cord blood

Now try this

How are stem cells used to make a new organ? (3 marks)

iPS could be a good source of stem cells in the future.

Exam skills

This question is about cell division and uses knowledge and skills that you have already revised. Look at pages 41–43, 46 and 47 to remind yourself about the cell cycle, mitosis, meiosis and stem cells.

Worked example

(a) What are the four stages of mitosis (in order)?
(1 mark)

prophase, metaphase, anaphase, telophase

> Make up a mnemonic to help you remember the order of the stages.

(b) Explain what happens during anaphase in mitosis. **(2 marks)**

The centromere of each chromosome splits. Sister chromatids are pulled apart to opposite sides of the cell.

> The sister chromatids are attached to a motor protein which drags them along the microtubule/spindle fibre to the pole of the cell.

(c) Draw the cells in anaphase 1 in meiosis (the parent cell had four chromosomes). **(2 marks)**

> There are four chromosomes, two homologous pairs. One half of each homologous pair should be on the opposite side of the cell. You can indicate this by shading. You should show each chromosome attached to a spindle fibre.

(d) Explain why a cell will not go through mitosis if there is any DNA damage. **(3 marks)**

There is a checkpoint in the cell cycle between the G_2 and M phase. If there is DNA damage it is repaired or the cell is not allowed to go through mitosis. This is to prevent uncontrolled cell division that could lead to a tumour.

> There is also another checkpoint between the G_1 and S phases, to make sure that there are no mistakes in the DNA before it is duplicated. If there are mistakes and the DNA is duplicated, it could lead to mutations.
>
> Mutations in genes involved with cell cycle control can lead to uncontrolled cell division.

(e) What are the differences between a somatic cell and a stem cell? **(2 marks)**

Stem cells have no Hayflick limit and can keep on dividing indefinitely but somatic cells have a limited number of divisions. Stem cells are undifferentiated and can differentiate into different cell types but somatic cells are fully differentiated.

> All somatic cells have a set Hayflick limit. They can divide a set number of times and then they go through programmed cell death (apoptosis).

> Embryonic stem cells are pluripotent and can develop into any somatic cell. Adult stem cells are multipotent and can divide into a limited number of cell types.

(f) What are some of the potential uses of stem cells in research and medicine? **(3 marks)**

To study normal cell development, to use as models for the study of diseases and to make replacement tissues and organs.

> Stem cells can divide indefinitely, so you can study them for many years in a lab.
>
> Some people have already received tissues and organs made from their own stem cells.

Gas exchange surfaces

Living organisms need to be able take in the oxygen they need for aerobic respiration. They also need to remove carbon dioxide waste.

Surface area to volume ratio

In **single cell** organisms, such as bacteria, this can be done by diffusion, without the need for a specialised exchange surface. This is because a bacterium's surface area to volume ratio (SA:V) is large. It has a large surface area compared to its volume.

In **multicellular** organisms, such as us, the SA:V ratio is very small. The oxygen could diffuse into our bodies but it would take too long to reach the cells. For this reason, we and other multicellular organisms need a specialised gas exchange surface to increase the SA:V ratio and maintain metabolic activity.

sides = 1
surface = $1^2 \times 6 = 6$
sides = $1^3 = 1$

sides = 3
surface = $3^2 \times 6 = 54$
sides = $3^3 = 27$

$\dfrac{\text{surface}}{\text{volume}} = 6$　　$\dfrac{\text{surface}}{\text{volume}} = 2$

Features of the gas exchange surface

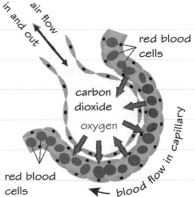

air flow in and out

red blood cells

carbon dioxide

oxygen

red blood cells

blood flow in capillary

In mammals, the gas exchange surface is in the lungs. The trachea branches into two bronchi, which in turn branch into many bronchioles. At the end of the bronchioles are air sacs called alveoli.

The features that make the lungs an efficient exchange surface are:

- Increased surface area due to many alveoli.
- The walls of the alveoli and the capillaries are thin (only one cell thick), decreasing the distance needed for the diffusion of gases.
- A good blood supply. There is a capillary close to each alveolus, rapidly taking the oxygenated blood away and bringing in deoxygenated blood. This ensures that a steep concentration gradient for oxygen and carbon dioxide is maintained.

Maths skills To work out the SA:V ratio of a cube, you need to know the surface area and the volume.

You then divide the surface area by the volume. For example, if the cube is 2 cm by 2 cm and has 6 sides, the surface area is:

$2 \times 2 \times 6 = 24\,\text{cm}^2$

The volume will be:

$2\,\text{cm} \times 2\,\text{cm} \times 2\,\text{cm} = 8\,\text{cm}^3$

That makes the SA:V ratio:

$\dfrac{24}{8} = 3$

The SA:V ratio is 3.

The histology of exchange surfaces

You can see that the walls of the alveoli are very thin. These are made from squamous epithelium (see page 45 for a reminder about this specialised tissue). The alveolar walls also contain some elastic fibres so that the alveoli can **recoil** back to their original size after expiration. The alveoli are associated with many capillaries.

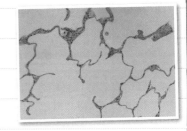

The gas exchange system in bony fish is an example (see page 53).

In plants, the root hair cells are an efficient exchange surface. They have a large surface area to increase the rate of diffusion.

Worked example

Why do multicellular organisms need a specialised gas exchange surface?　　**(3 marks)**

They have a small SA:V ratio.

Gases will not diffuse quickly enough.

Specialised gas exchange surfaces increase the surface area, and therefore the SA:V ratio.

Now try this

Why is it important to maintain a steep concentration gradient for the gases being exchanged?　　**(2 marks)**

49

The lungs

Structure of the lungs

Air enters the lungs through the nose or mouth and goes down the **trachea**. The trachea branches into two **bronchi** (the right and left bronchus) in the lungs. Each bronchus branches into smaller and smaller **bronchioles**, eventually ending at small air sacs called **alveoli** (singular alveolus).

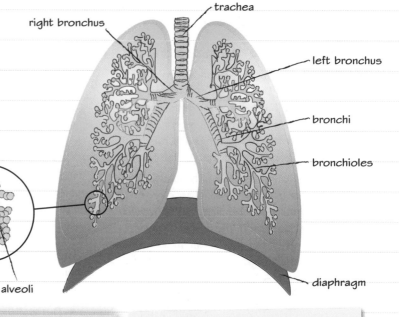

right bronchus

trachea

left bronchus

bronchi

bronchioles

diaphragm

alveoli

Worked example

What is the purpose of the elastic fibres in the alveoli?   **(2 marks)**

Alveoli stretch during inspiration.

Elastic fibres allow alveoli to recoil after expiration.

The elastic fibres in the trachea, bronchi and bronchioles have a different role; they recoil and allow the lumen of these airways to go back to original size after being constricted.

Look at page 51, for more about the mechanism of ventilation.

Tissues of the lungs and their distribution and function

Tissue	Histology	Distribution	Function
Cartilage		Trachea, bronchi	Holds the airways open
Goblet cells		Trachea, bronchi	Secrete mucus in order to trap pathogens and particles in the lungs
Smooth muscle		Trachea, bronchi, bronchioles	Constricts airways to prevent harmful gases from entering the lungs
Ciliated epithelium		Trachea, bronchi, bronchioles	Surface is covered in cilia that move mucus out of the lungs
Elastic fibres		Trachea, bronchi, bronchioles, alveoli	Allow recoil of the smooth muscle or the alveoli back to their original size

Now try this

1 Heavy smokers often lose the cilia in their trachea and bronchi. What effect might this have?
   (2 marks)

2 Suggest the role of the smooth muscles in the symptoms of asthma.
   (3 marks)

The mechanism of ventilation

Breathing is divided into two actions, **inspiration** (breathing in) and **expiration** (breathing out).

Inspiration

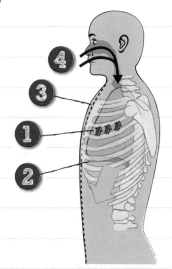

Expiration

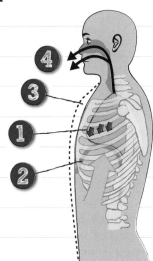

1 Intercostal muscles contract. The ribcage moves upwards and outwards.

2 Diaphragm contracts and becomes flatter.

3 Thoracic cavity expands and volume increases. This decreases the pressure to below atmospheric pressure.

4 Air moves into the lungs from high pressure to low pressure.

1 Intercostal muscles relax. The ribcage moves downwards and inwards.

2 Diaphragm relaxes and becomes dome-shaped.

3 Thoracic cavity reduces and volume decreases. This increases the pressure above atmospheric pressure.

4 Air moves out of the lungs from high pressure to low pressure. The recoil of the elastic fibres in the alveoli helps to expel this air.

Worked example

Why do the ribcage and the diaphragm have to move during expiration? **(3 marks)**

To decrease the volume of the thorax.

To increase the pressure inside the thorax.

So that air will move from high pressure inside the lungs to low pressure outside of the lungs.

Remember that the ribcage and diaphragm move because of the contraction and relaxation of muscles.

Air movement is always to do with changes in pressure.

Now try this

1 Hiccups occur when the diaphragm contracts involuntarily. Explain what will happen next. **(2 marks)**

2 At high altitude, e.g. on top of a mountain, it is more difficult to breathe. Why do you think this is? **(3 marks)**

Using a spirometer

Practical skills A spirometer is used to measure air movements in and out of the lungs.

A spirometer

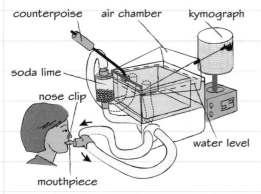

The air chamber is filled with air or oxygen. When a person exhales into the spirometer, the air chamber rises. When a person inhales, air is removed from the spirometer and the air chamber moves down. This movement draws the graph. The **soda lime** absorbs any carbon dioxide in the exhaled air.

A spirometer trace can also be used to work out a person's breathing rate and rate of oxygen uptake.

A spirometer trace

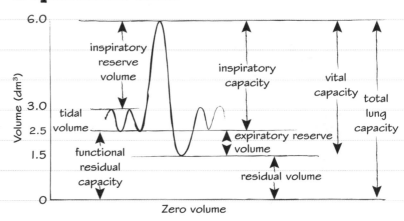

The **tidal volume** is the volume of air breathed in and out of the lungs in one breath, when the person is at rest. To measure this, measure the distance from the top to the bottom of a small peak. On this graph, the tidal volume is $0.5\,dm^3$.

The **vital capacity** is the maximum volume of air that can be inhaled or exhaled in one breath. To measure this, measure the distance from the top to the bottom of the largest peak. On this graph, the vital capacity is $4.5\,dm^3$.

Remember: Inhale = breath in   Exhale = breath out

Data from a spirometer

At rest, the tidal volume is small, usually around $0.5\,dm^3$, and the average breathing rate is 12–15 breaths per minute. The rate of oxygen uptake is also low.

As you exercise, the tidal volume, breathing rate and rate of oxygen uptake all increase.

If you exercise regularly, your resting vital capacity will be much higher.

Maths skills ## How to work out breathing rate and rate of oxygen uptake

Breathing rate: divide the number of breaths by a given time, and multiply by 60 to get the breaths per minute. For example, if there are 11 breaths in 30 seconds:

$$\frac{11}{30} \times 60 = 22 \text{ breaths per minute}$$

Rate of oxygen uptake: divide the amount of oxygen used by a given time and multiply by 60 to get the oxygen uptake per minute. For example, $0.38\,dm^3$ used in 30 seconds:

$$\frac{0.38}{30} \times 60 = 0.76\,dm^3 \text{ per minute}$$

Worked example

A person breathes 12 times in 36 seconds. What is their breathing rate? **(2 marks)**

$$\frac{12}{36} \times 60 = 20 \text{ breaths per minute}$$

Before using a spirometer, the equipment must be sterilised. A person who is feeling ill should not use it.

Now try this

A person uses $1.8\,dm^3$ of oxygen in 45 seconds. What is their rate of oxygen uptake? **(2 marks)**

Ventilation in bony fish and insects

Bony fish and insects have different ventilation systems to mammals.

Ventilation in bony fish

Bony fish use gills to absorb oxygen dissolved in the surrounding water.

Water enters the fish's mouth, or **buccal cavity**, when the mouth opens, lowering the pressure inside the buccal cavity compared to the surrounding water. When the fish closes its mouth, the pressure in the buccal cavity increases and water flows over the gills and out of the fish, behind a flap called the **operculum**.

The **gills** are made of a series of v-shaped **lamellae** through which blood flows. As the water flows over the deoxygenated blood, oxygen diffuses into the blood. The blood flows in the opposite direction to the water. This is called the **countercurrent mechanism**. This is important because it allows the maximum amount of oxygen to diffuse into the blood.

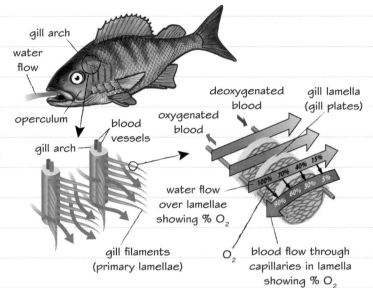

Ventilation in insects

Insects use **spiracles** to take in oxygen from the air through the sides of their bodies.

Insects have an open circulatory system, which means that the blood flows out of the blood vessels and into the body cavity. Oxygen from the air diffuses into the insect's body cavity through the spiracles, and into long thin tubes called **tracheae**. The tracheae branch into smaller tubes called **tracheoles** that have an open ending inside the insect cell, filled with **tracheal fluid**. Oxygen diffuses into this fluid, and into the insect's cells. The rate of diffusion can be increased in active insects by withdrawing the tracheal fluid into the cells. This increases the surface area of the tracheole wall that is exposed to air, and more oxygen can be absorbed. Some very active insects can help this ventilation system further by increasing and decreasing the volume of the thorax by moving their wings.

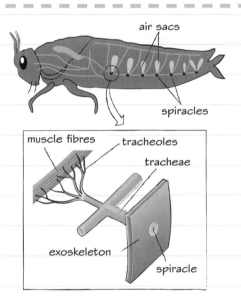

Worked example

Label the gills of a fish as seen on this histology slide.

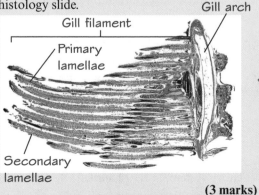

(3 marks)

Now try this

1 Explain how the countercurrent mechanism works. **(3 marks)**
2 How do insects increase the amount of oxygen taken in through the spiracles? **(2 marks)**

You will be expected to do a dissection of a bony fish and/or an insect to observe and draw the gaseous exchange system.

Circulatory systems

All multicellular organisms need a circulatory system in order to make sure that oxygen and metabolites reach all of the tissues, and so that the waste products, such as carbon dioxide, are removed. There are several different types of circulatory system.

The need for a circulatory system

Small organisms, such as bacteria, do not need a circulatory system because they can receive oxygen and nutrients by diffusion alone. The larger the size and metabolic rate of the organism, and the smaller the SA : V ratio, the more complex the circulatory system becomes. (Look back at page 49.)

Simpler, smaller organisms can use an **open circulatory system**. Larger, more active organisms use a **closed circulatory system**.

Open circulatory system

Insects often use an open circulatory system.

The blood does not stay inside the blood vessels, but can leave and go into the body cavity. Blood in the body cavity re-enters the blood vessels through pores called **ostia**, where it is pumped by a single-chambered heart back into the body cavity.

anterior vessel

lateral vessels

tubular heart

ostia

flow of blood into the body cavity

blood passes back into the heart

Closed circulatory systems

Single and double circulatory systems are both closed circulatory systems. The blood stays inside blood vessels at all times. This improves the speed at which oxygen and nutrients can reach all of the cells.

Fish have a **single circulatory system**. The blood goes through the heart once. The blood is pumped through a two-chambered heart, across the gills and around the body, and back to the heart.

Mammals have a **double circulatory system**. The blood goes through the heart twice.

The blood goes through the two chambers on the right side of the heart and then to the lungs to collect oxygen. The blood then returns to the left side of the heart and travels around the body and back to the right side of the heart. This system allows the blood to travel at a lower pressure through the delicate capillaries of the lungs, and at a higher pressure once out of the lungs to get quickly around the body.

capillary beds in tissue

veins returning to the heart

arteries to head

pulmonary artery carrying blood to the lungs

arteries to the upper body and head

pulmonary circulation

lungs

pulmonary vein carrying blood back to the heart

right atrium

left atrium

left ventricle

right ventricle

systemic circulation

heart – the pump which forces blood around the body

aorta – the major artery leaving the heart

■ deoxygenated blood
■ oxygenated blood

capillary beds in tissue

arteries to the body

Double circualtory system

Insects are different

Insects have a one-chamber heart, whereas bony fish have a two-chamber heart. Mammalian hearts have four chambers.

Insect blood is different to fish and mammal blood and is called **haemolymph**.

Worked example

Compare an open and a closed circulatory system.
(2 marks)

Both systems have a heart. In the open circulatory system, blood leaves the blood vessels, but in a closed circulatory system, the blood stays inside the blood vessels.

Now try this

Why do larger, more metabolically active organisms need a closed circulatory system?   **(2 marks)**

Blood vessels

There are five different types of blood vessel in the human body: arteries, arterioles, capillaries, venules and veins. They share common tissues but have different structures.

Tissues of the arteries, veins and capillaries

Lumen – the space inside the blood vessel where the blood flows.

Endothelium – the inside smooth layer of cells made of **squamous endothelium**.

Elastic fibres – allow blood vessel to **recoil** to its original size after being stretched or constricted.

Smooth muscle – can contract to constrict or narrow the blood vessel.

Collagen fibres – allow the blood vessels to withstand the pressure of the blood.

Valves – prevent the backflow of blood.

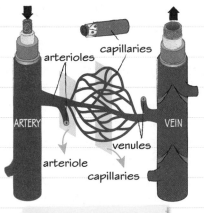

Structure of artery, vein and capillary wall

Properties of arteries

- small lumen to **maintain** a high pressure
- highly folded endothelium, so lumen can get wider as blood passes through
- wall is very thick, to **withstand** the high pressure
- elastic fibres recoil after being stretched by the flow of blood (the **pulse**).

Arteries branch into smaller **arterioles**. These have a thick layer of smooth muscle that can be used to constrict the arteriole and divert blood away from an area.

Properties of veins

- wide lumen, as the blood pressure is much lower in the veins
- presence of **valves**
- wall is very thin as it does not need to withstand a high pressure
- has a thin layer of elastic fibres as the veins do not stretch and recoil
- flow is helped by the contraction of the skeletal muscles.

Veins branch into smaller **venules**. These have a thin layer of smooth muscle, elastic fibres and collagen fibres.

Worked example

Suggest two occasions when blood would be diverted in the body by arteriole constriction. **(2 marks)**

In thermoregulation, arterioles close to the skin surface are constricted to divert blood flow from the skin surface and reduce heat loss.

During the fight or flight response, blood is diverted to the skeletal muscles and away from the digestive system.

Properties of capillaries

- consist of lumen and a layer of endothelium
- endothelial cells have small gaps between them to allow the movement of cells such as **neutrophils** into the tissues.

Heat can be lost from the skin surface by radiation.

The skeletal muscles require more blood so that they can increase aerobic respiration.

Now try this

1 Describe three ways in which arteries and veins are different.   **(3 marks)**

2 Explain how arteries maintain a high pressure.   **(2 marks)**

The formation of tissue fluid

Tissue fluid is made from the blood plasma. It has a similar composition to blood, but is missing erythrocytes and large proteins.

The formation of tissue fluid

- At the arteriole end of the capillary water moves out of the plasma and into the tissue fluid due to the overall pressure difference.

- When water moves out of the blood plasma, it takes dissolved molecules with it. Large molecules, such as proteins, and erythrocytes are too large to move to out of the capillary.

- Neutrophils can move into the tissue fluid through the small gaps between the cells of the capillary endothelium.

- At the venous end of the capillary, the water moves out of the tissue fluid and back into the plasma, again due to the overall pressure difference.

Excess water, excreted proteins and lipids leave the tissue fluid and flow into **lymph vessels**. This fluid forms the **lymph**. Lymph vessels contain pores to allow the entry of large molecules, such as proteins. Lymphocytes, a type of white blood cell, can be found in the lymph.

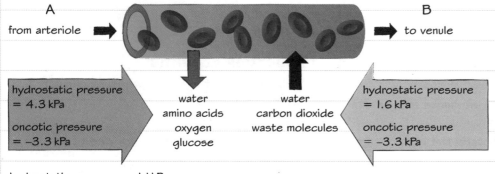

A   from arteriole →

B   → to venule

hydrostatic pressure = 4.3 kPa

oncotic pressure = −3.3 kPa

water
amino acids
oxygen
glucose

water
carbon dioxide
waste molecules

hydrostatic pressure = 1.6 kPa

oncotic pressure = −3.3 kPa

hydrostatic pressure = 1.1 kPa
oncotic pressure = −1.3 kPa
overall pressure = 1.2 kPa outwards

hydrostatic pressure = 1.1 kPa
oncotic pressure = −1.3 kPa
overall pressure = 1.5 kPa inwards

Hydrostatic pressure: the pressure that a fluid exerts when pushing against the sides of a vessel or container.

Oncotic pressure: the pressure created by the solutes.

Components of blood, tissue fluid and lymph

Blood	Tissue fluid	Lymph
Erythrocytes, neutrophils and lymphocytes	Neutrophils	Lymphocytes
Large blood proteins	Few excreted proteins	Few excreted proteins
High oxygen	Low oxygen	Low oxygen
High glucose	Low glucose	Low glucose
High amino acids	Low amino acids	Low amino acids
Low carbon dioxide	High carbon dioxide	High carbon dioxide
Lipids	Few lipids	Many lipids

Now try this

Why does blood plasma have a lower water potential than the tissue fluid?   **(2 marks)**

Worked example

Why does water from the blood plasma move into the tissue fluid at the arterial end but not the venous end of the capillary?   **(3 marks)**

High hydrostatic pressure at arteriole end.

More than oncotic pressure, so water moves into tissue fluid.

Oncotic pressure greater than hydrostatic pressure at venous end.

Sometimes, as here, hydrostatic pressure and oncotic pressure are demonstrated on diagrams using figures. You will **not** be asked to make calculations using these figures.

The mammalian heart

The mammalian heart is part of a double circulatory system. Blood flows through the right side of the heart to the lungs and through the left side of the heart to the rest of the body.

External and internal structure

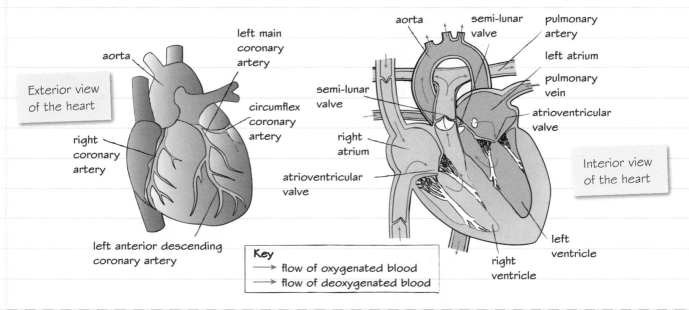

Exterior view of the heart

aorta

left main coronary artery

circumflex coronary artery

right coronary artery

left anterior descending coronary artery

aorta

semi-lunar valve

pulmonary artery

left atrium

pulmonary vein

atrioventricular valve

semi-lunar valve

right atrium

atrioventricular valve

right ventricle

left ventricle

Interior view of the heart

Key
→ flow of oxygenated blood
→ flow of deoxygenated blood

Blood flow

Deoxygenated blood from the body flows:

1. into the heart through the superior and inferior vena cava
2. into the right atrium
3. through the atrioventricular valve into the right ventricle
4. out of the right side of the heart through the semi-lunar valves and into the pulmonary artery then to the lungs, where it is oxygenated.

Oxygenated blood from the lungs flows:

1. back into the heart through the pulmonary vein
2. into the left atrium
3. through the atrioventricular valve into the left ventricle
4. out of the left side of the heart through the semi-lunar valves, into the aorta and around the body.

Dissection

You will be expected to carry out a dissection of the heart, identify all of the blood vessels, chambers and valves and draw them. It is necessary to learn all of the labels of the heart.

Points to remember

- **cardiac** = related to the heart
- **pulmonary** = related to the lungs.
- Arteries carry blood **A**way from the heart.
- Veins carry blood towards the heart.

Now try this

Compare and contrast the blood in the pulmonary artery and the aorta. **(2 marks)**

Worked example

Explain why the left ventricle appears to be larger than the right ventricle. **(3 marks)**

The left ventricle has a thicker muscle wall in order to create a higher pressure, and pump blood around the whole body.

Left and right are with respect to the body that contains the heart, looking forwards, remember this when you look at a diagram of a heart.

The cardiac cycle

Blood flows through the heart because of a series of pressure changes called the **cardiac cycle**.

① Diastole

- Both the atria and the ventricles relax.
- The pressure in the arteries is higher than the pressure in the ventricles, so the semi-lunar valves close.
- Blood flows into the atria.

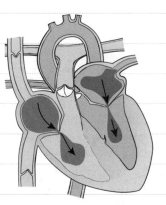

② Atrial systole

- Both of the **atria** contract at the same time.
- The pressure inside the atria is higher than the pressure in the ventricles.
- Blood flows through the open **atrioventricular valves** (AV valves) into the ventricles.

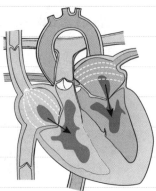

③ Ventricular systole

- Both of the **ventricles** contract at the same time.
- Pressure inside the ventricles is higher than pressure in the atria and the arteries leaving the heart, causing the AV valves to close and the **semi-lunar valves** in the arteries to open.
- Blood leaves the heart.

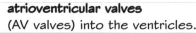

Blood pressure graph

Blood pressure is measured with a **sphygmomanometer** and is recorded as two numbers.

- The first number is the **systolic pressure** (arterial pressure when the ventricles contract).
- The bottom number is the **diastolic pressure** (arterial pressure when the ventricles are relaxed).

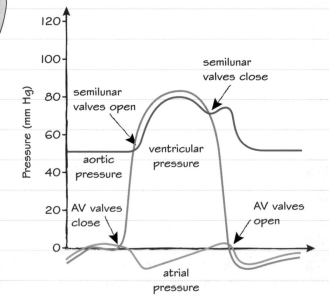

Worked example

Explain why the atrioventricular valves close during ventricular systole. **(2 marks)**

The pressure is higher in the ventricles than in the atria. The blood pushes the atrioventricular values shut and the tendinous cords stop them from opening inside out.

The atrioventricular valves can be referred to as the AV valves, but always use their full name at least once in your answer.

Now try this

1   Describe the shape of the graph for the ventricles.   **(3 marks)**
2   Why does the pressure in the left ventricle have to exceed the pressure in the aorta?   **(2 marks)**

Control of the heart

The regular beat of the heart at rest is controlled by the heart's natural pacemaker, the sino-atrial node (SAN).

Electrical conductivity of the heart

The resting heart rate is usually between 55 and 80 beats per minute. The cardiac muscle of the heart is **myogenic**. This means that the cells conduct electricity and can initiate their own contractions.

The heartbeat is initiated by the **sino-atrial node** (SAN) at the top of the right atrium. The SAN sends out an electrical **wave of excitation** across both of the atria, causing them to contract at the same time. At the top of the septum is the **atrioventricular node** (AVN). This sends the wave of excitation along the **purkyne** tissue in the septum to the **apex** (bottom of the heart). This causes the ventricles to contract from the apex upwards. There is a slight delay of 0.1 milliseconds (ms) between the contractions of the atria and the ventricles, to allow the atria to empty.

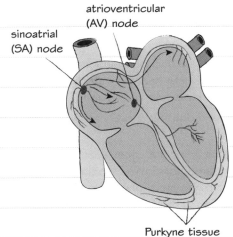

atrioventricular (AV) node

sinoatrial (SA) node

Purkyne tissue

Normal electrocardiogram (ECG)

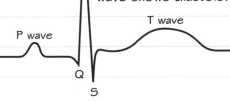

R This is a normal electrocardiogram. The P wave shows the atrial systole, the QRS complex shows the ventricular systole and the T wave shows diastole.

P wave

T wave

Q

S

QRS complex

Terms to remember

Atrial systole — the contraction of the atria.

Ventricular systole — the contraction of the ventricles.

Diastole — relaxation of the atria and ventricles.

Interpreting ECGs

🖩 Maths skills

ECGs can be used to diagnose heart problems.

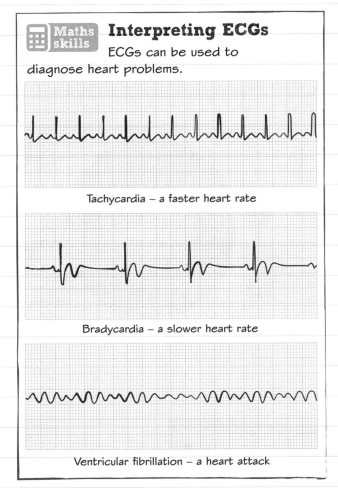

Tachycardia – a faster heart rate

Bradycardia – a slower heart rate

Ventricular fibrillation – a heart attack

Why is the QRS complex larger than the P wave on a normal ECG? **(2 marks)**

The QRS complex represents the wave of excitation across the ventricles and the P wave represents the wave of excitation across the atria. The ventricles are larger and have a thicker muscle wall than the atria, so need a larger wave of excitation in order to contract.

The muscle walls of the heart chambers contract after the wave of excitation passes over them.

1 Explain why it is important that the ventricles contract from the apex upwards. **(2 marks)**

2 Arrhythmia is an irregular heartbeat. Suggest how the ECG of this condition would appear. **(2 marks)**

Haemoglobin

Haemoglobin carries oxygen and carbon dioxide around the body.

Haemoglobin and oxygen

Haemoglobin is a globular protein made of four subunits. See page 17 for an illustration.

Each subunit contains an iron-containing **haem** group that can carry one molecule of oxygen (O_2):

Haemoglobin + $4O_2$ ⇌ Oxyhaemoglobin
$$(Hb4O_2)$$

This is a reversible reaction. Haemoglobin will associate with O_2 at high partial pressure of O_2, and dissociate with O_2 at low partial pressure of O_2.

🖩 Maths skills | The oxygen dissociation curve

The fetal haemoglobin curve is to the left of the normal (adult) haemoglobin oxygen dissociation curve. The Bohr shift moves the normal curve to the right. In actively respiring cells, more CO_2 is released. This causes more H^+ ions to be present in the red blood cells, and for more O_2 to be released to the active tissues.

Haemoglobin and carbon dioxide

Of the carbon dioxide (CO_2) carried in the blood, 85% is carried in the red blood cells as hydrogen carbonate ions (HCO_3^-), 5% is dissolved in the blood plasma and 10% is carried by haemoglobin as carbaminohaemoglobin ($HbCO_2$).

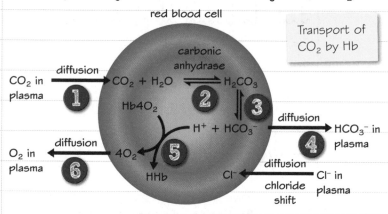

Transport of CO_2 by Hb

Partial pressure of oxygen is the amount of oxygen available. The lungs have a high ppO_2 and the tissues have a low ppO_2.

1 CO_2 diffuses into the RBC.

2 Carbonic anhydrase enzyme forms H_2CO_3.

3 H_2CO_3 breaks down.

4 HCO_3^- diffuses out.

5 H^+ causes $Hb4O_2$ to release O_2.

6 O_2 diffuses out of RBC.

The HCO_3^- ions diffuse out of the red blood cell into the blood plasma. This unbalances the charge of the red blood cell so Cl^- ions diffuse into it to redress the balance. This is called the **chloride shift**.

Worked example

Explain the differences between the oxygen dissociation curves of a fetus and its mother. **(3 marks)**

Fetal haemoglobin has a higher affinity for oxygen than adult haemoglobin.

At low partial pressures of oxygen (ppO_2), oxygen dissociates from adult haemoglobin and associates instead with fetal haemoglobin.

Affinity means the strength of binding to a substance. Haemoglobin has a high affinity for oxygen.

When oxygen binds to haemoglobin, we say it **associates**. When oxygen is released from haemoglobin it **dissociates**.

Now try this

1 How many molecules of oxygen can be carried by one haemoglobin molecule? **(1 mark)**

2 Why is the Bohr shift useful in active tissues? **(2 marks)**

The plant vascular system

Multicellular plants are large and have a small surface area to volume ratio. They could not transport substances by diffusion alone, so they need a vascular system to transport water, minerals and sugars around the plant.

Xylem

Xylem tissue is a hollow, narrow tube that transports water and minerals from the roots to the leaves of the plant.

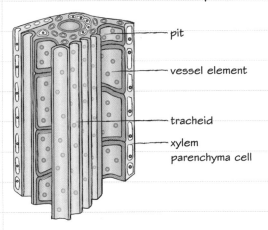

pit

vessel element

tracheid

xylem parenchyma cell

Phloem

Phloem tissue transports sugars and other assimilates up and down the plant.

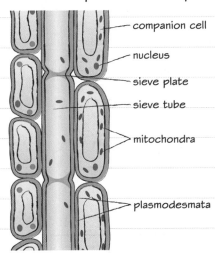

companion cell

nucleus

sieve plate

sieve tube

mitochondra

plasmodesmata

- Xylem is made of cells called **xylem vessel elements**, stacked end to end.
- Xylem cells are dead, with no cytoplasm or organelles and no end plates between the cells.
- The walls of the xylem contain **lignin**, which adds strength to the xylem and keeps water inside.
- Water can pass to nearby xylem through **pits** in the xylem wall.

- Phloem is made of cells called **sieve tube elements**, stacked end to end to make a narrow tube, with a sieve plate between them.
- Sieve plates contain many pores that allow sugars through.
- Phloem cells are living, but only contain a thin layer of cytoplasm around the edge of the cell and few organelles.
- Each phloem cell is supported by a **companion cell** that loads sugars (mostly **sucrose**) into the phloem. Companion and phloem cells are linked by pores in the phloem wall called **plasmodesmata**.

Worked example

How is the xylem adapted to transport water?

(2 marks)

Xylem tissue forms hollow, narrow tubes, for the transport of water.

Xylem cell walls contain lignin for strength and to waterproof the xylem.

> You will only be asked about the vascular systems of herbaceous dicotyledonous plants. These are a category of plants that form two leaves when they germinate from a seed.

Now try this

1 What is the role of the companion cells in the phloem tissue? **(2 marks)**
2 Name three structural differences between xylem and phloem tissue. **(3 marks)**

Leaves, stems and roots

Leaves, stems and roots each have a different arrangement of xylem and phloem. You can see this when dissecting and examining plant tissue.

The distribution of xylem and phloem

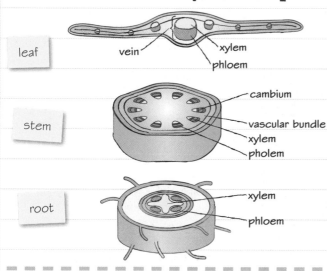

leaf — vein, xylem, phloem

stem — cambium, vascular bundle, xylem, pholem

root — xylem, phloem

- Leaves: xylem and phloem are arranged in veins; xylem is towards the top and the phloem is towards the bottom.
- Stems: xylem and phloem are arranged into **vascular bundles** around the outside, with the xylem on the inside and the phloem on the outside and a layer of stem cells (**cambium**) between them.
- Roots: xylem in the centre, usually in an 'x' shape, with the phloem around the xylem.

Examination of stained plant tissue

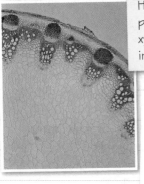

Here we can see the phloem (red) and the xylem (brown) arranged into vascular bundles.

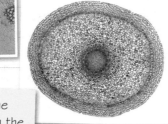

Here we can see the xylem and phloem in the centre of the root.

Xylem in dissected plant stems

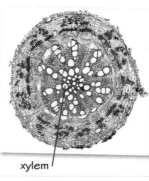

We can see the lignin of xylem as large red circles.

xylem

Here we can see the lignin of xylem as red lines.

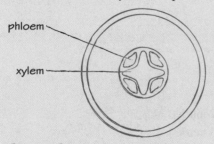

Now try this

1 What features of xylem are observable in the longitudinal dissection?   **(2 marks)**
2 What is the role of the cambium?   **(2 marks)**

Transpiration

Transpiration is the loss of water vapour from the upper parts of the plant. It is a consequence of gaseous exchange.

Transpiration from the leaves

1 Water leaves the xylem by **osmosis** and enters the **mesophyll cells**.

2 Water continues moving through the mesophyll cells to the palisade cells by osmosis, to be used in photosynthesis.

3 Some of the water **evaporates** from the mesophyll cells and forms **water vapour** inside the air spaces of the leaf.

4 The **water vapour potential** is higher inside the leaf, than outside the leaf, so the water vapour leaves the underside of the leaf by **diffusion**, through pores called the **stomata**.

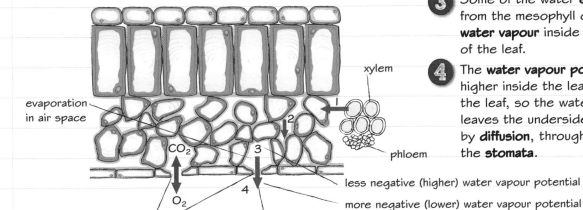

evaporation in air space

xylem

phloem

CO_2

O_2

less negative (higher) water vapour potential

more negative (lower) water vapour potential

stomata open to allow exchange of gases

water vapour lost by diffusion down water vapour potential gradient into surrounding air

Environmental factors and transpiration

Factors that increase transpiration	Factors that decrease transpiration
Increase in temperature	Decrease in temperature
Increase in sunlight	Decrease in sunlight
Increase in wind movement	Decrease in wind movement
Decrease in humidity	Increase in humidity
More stomata open	Fewer stomata open

It is important not to let any air bubbles into the xylem of the plant. This is done by cutting the stem and preparing the plant in the potometer under water.

Estimating transpiration rates

Transpiration rates can be estimated using a **potometer**. An air bubble is introduced into the capillary tube and the distance moved by the bubble over time gives a measure of the amount of water taken up by the plant. This is used to indicate how much water has been lost by transpiration.

This is a reasonably accurate measure, as 95% of the water is lost by transpiration.

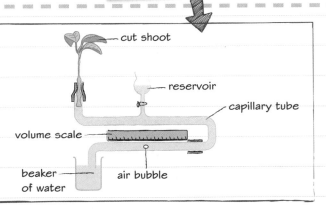

cut shoot

reservoir

capillary tube

volume scale

beaker of water

air bubble

Now try this

1 Explain why more stomata open when it is light. **(2 marks)**
2 Design an experiment to measure the rate of transpiration at different light levels.
   (5 marks)

The transport of water

The pathways of water through the roots

Water enters the **root hair cells** by osmosis. It moves by osmosis across the **cortex**, through the **endodermis**, to the xylem. Minerals are actively transported across the endodermis and into the xylem, lowering the water potential in the xylem.

There are three pathways that water can take:

1 **symplastic** pathway — water moves through the cytoplasm of adjacent cells

2 **apoplastic** pathway — water moves through the spaces between the cell wall and plasma membrane of adjacent cells, and in the spaces between cells

3 **vacuolar** pathway — water moves through the cytoplasm and vacuoles of adjacent cells.

Water cannot cross the waterproof **suberin** in the **Casparian strip**. Therefore, all water enters the xylem through the symplast pathway.

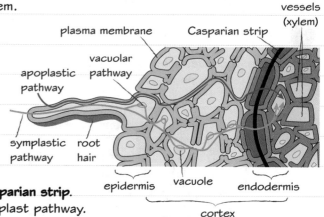

Movement of water up the stem

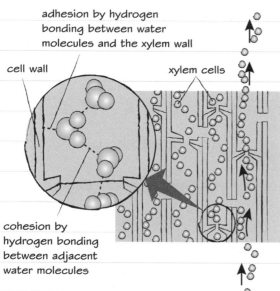

Water moves up the xylem in a **transpiration stream** held together by cohesion and adhesion.

This creates a tension and is known as the **cohesion–tension theory**. Water is pushed up the xylem by the **root pressure** of water, and pulled up the xylem by transpiration.

Adaptations

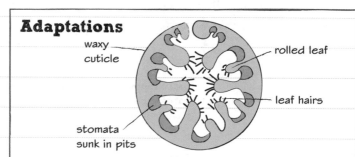

Xerophytes are plants that are adapted to live in habitats with little available water – like deserts and the extreme cold. Examples are cacti and marram grass.

Xerophyte adaptations:

- conserve water by storing water in stems or leaves
- prevent water being lost by having coiled leaves, sunken stomata and small hairs around the stomata to catch water droplets. This increases the **water vapour potential** around the stomata and lowers the water vapour potential gradient, decreasing transpiration.

Hydrophytes are plants that live in water, for example, water lilies.

Hydrophyte adaptations:

- many large air spaces in the leaf to keep leaves afloat
- the stomata are on the upper epidermis, so that they are exposed to the air. This allows gaseous exchange from the surface of the leaves and the transpiration of water.

Worked example

Explain how water moves from the root epidermis into the xylem. **(3 marks)**

Minerals are actively transported across the endodermis into the xylem. This decreases the water potential in the xylem. Water moves into the xylem by osmosis. In turn, this creates a water potential gradient across the cortex.

Now try this

Explain how xerophyte adaptations reduce the rate of transpiration. **(4 marks)**

Translocation

Translocation is the energy-requiring movement of sugars and other assimilates up and down the plant.

The mass flow theory of translocation

Phloem sap contains sugars, mainly sucrose, and other assimilates, such as amino acids.

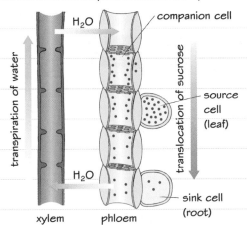

1 The sap is loaded onto the top of the phloem by **active loading** using ATP made by companion cells. This decreases the water potential of the sap.

2 Water from the xylem moves into the phloem by **osmosis**, creating pressure. This pressure forces the sap along the phloem.

3 At the other end of the phloem, the components of the sap are actively loaded out, increasing the water potential in the phloem. Water moves back into the xylem, and is transported to the top of the plant.

Sources and sinks

In the summer, excess sugars are made in the leaves by **photosynthesis**. These sugars are stored as **starch** in the roots. The leaves are the **source** and the roots are the **sink**.

In the spring, sugars are needed in the buds, flowers and fruits. These are used in respiration to make the ATP needed for growth. The roots are then the source and the buds, flowers and fruits are the sink.

Practical skills — Evidence for the mass flow theory

This experiment shows evidence for the mass flow theory:

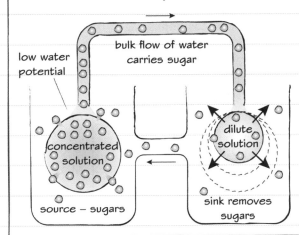

Water moves across the partially permeable membrane into the high concentration sugar solution flask, creating a pressure. The solution then moves through the tube into the other flask.

Term to remember

Assimilates are substances that have become part of the plant. They are made by the plant, mostly from glucose, and include other sugars, such as sucrose, lipids and amino acids.

Worked example

What is the role of the water in mass flow? **(3 marks)**

Water moves out of the xylem into the phloem.

Water potential is lower in the phloem than the xylem.

Pressure of water at the top of the phloem forces the sap down the phloem.

Now try this

1 What is meant by a source and a sink? **(2 marks)**

2 Explain how the experiment above supports the mass flow theory. **(3 marks)**

Exam skills

This question uses knowledge and skills that you have already revised. Look at pages 61–65 to remind yourself about transport in plants.

Worked example

(a) Give two features of the xylem that allow it to transport water. **(2 marks)**

hollow tube and waterproof

Xylem tissue is hollow because the xylem vessel element cells have no end plates.

Lignin in the xylem cell walls makes the xylem waterproof, although some water can move horizontally through pits (gaps in the lignin).

(b) What is the name of the process that moves sucrose and other assimilates through the plant? **(1 mark)**

translocation

(c) Draw and label the distribution of xylem and phloem in the stem of a dicotyledonous plant. **(2 marks)**

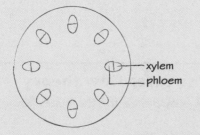
xylem
phloem

Xylem and phloem in the stems are arranged into vascular bundles around the edge of the stem. Xylem is always towards the centre, and phloem is always towards the outside of the stem.

(d) The rate of transpiration under different conditions was estimated using a potometer. The water level was 30 cm at the start of each experiment. Complete the table. **(2 marks)**

Conditions	Water level after 30 minutes (cm)	Rate of transpiration (cm/min)
Dark + warm	28	0.067
Light + warm	25	0.100
Dark + cold	30	0.000
Light + cold	29.5	0.017

Maths skills — Calculating the rate of transpiration

You need to know how much water (in cm) has been transpired and the time taken.

For the light + warm condition, 5 cm of water was transpired in 30 minutes. Therefore the rate of transpiration = $\frac{5}{30}$ = 0.100.

Make sure all of the rates are to the same number of decimal places.

(e) Explain why the rate of transpiration was fastest in the light and warm condition. **(2 marks)**

More stomata are open in the light. Water molecules have increased kinetic energy and are moving faster in warmer temperatures, so will diffuse out of the stomata more quickly.

Many plants close their stomata at night, and open them in the day when they are photosynthesising. This prevents too much water being lost by transpiration.

(f) Xerophytes have adaptations that reduce the rate of transpiration. Name two adaptations and explain how they work. **(4 marks)**

Sunken stomata are not exposed to the wind, and small hairs on the underside of the leaf trap water vapour. This means that the wind does not remove water vapour, which remains just outside the stomata, reducing the water vapour potential gradient.

A xerophyte is a plant that is adapted to dry conditions, e.g. a cactus.

Water vapour diffuses down the water vapour potential gradient from an area of high water vapour potential to an area of low water vapour potential. The higher the gradient, the faster the rate of diffusion.

Types of pathogens

A pathogen is a microorganism that causes disease. There are many different types of pathogens that can cause communicable diseases in plants and animals.

1 Bacteria

Bacteria are simple, single-celled organisms. Using the five-kingdom classification, they belong to a kingdom called **Prokaryotae**. Bacteria can divide and spread rapidly and cause diseases.

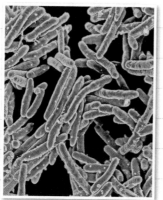
Tuberculosis

Diseases:

- tuberculosis (TB)
- bacterial meningitis
- ring rot (in potatoes and tomatoes).

2 Viruses

Viruses are much smaller than bacteria, and are not true organisms, since they cannot survive for long without a host.

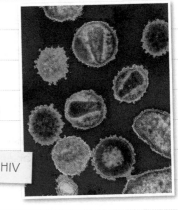
HIV

Diseases:

- AIDS caused by the human immunodeficiency virus (HIV) in humans
- influenza (in many animals)
- infection with tobacco mosaic virus (in plants).

3 Protoctista

Protoctista are single-celled organisms that are **eukaryotic** (they contain a nucleus). They are often carried by **vectors**, such as mosquitos (see page 68 for more about transmission of pathogens).

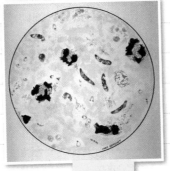
Malaria

Diseases:

- malaria
- potato blight / tomato late blight.

4 Fungi

Fungi can be single-celled or multicellular. They are a type of eukaryotic cell and have a nucleus, organelles and a chitin cell wall.

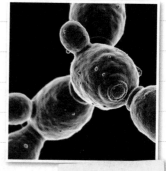
Black Sigatoka

Diseases:

- black sigatoka (in bananas)
- ringworm (in cattle)
- athlete's foot (in humans).

Worked example

Give one example of a plant disease that is caused by (a) a protoctista and (b) a fungus. **(2 marks)**

(a) Potato / tomato late blight is caused by a protoctista.

(b) Black sigatoka is caused by a fungus.

> These examples are mentioned in the specification, so should be used in the exam.

Now try this

1 Name four differences between a bacterium and a virus. **(4 marks)**
2 Suggest how malaria can be passed from human to human. **(2 marks)**

Transmission of pathogens

Pathogens can be transmitted in a number of different ways.

Direct transmission

- When mothers pass pathogens onto their unborn child, it is called **vertical transmission**. The pathogen moves from the mother's blood, across the placenta and into the child.
- **Horizontal transmission** is when a pathogen is passed between people by touching, kissing or sexual intercourse.
- Pathogens can be transmitted as droplets in the air when people cough or sneeze, as well as on food, in water or on spores.
- Plant pathogens can be transmitted through the soil or on spores carried on the wind.

These diseases are spread by direct transmission

Disease	Means of transmission
HIV	From mother to child; sexual intercourse; contact with infected blood.
Tuberculosis	Droplets in the air from coughing and sneezing.
Cholera	Drinking infected water.
Anthrax	Travels as spores through the air; spores present in the soil.

Indirect transmission

- Pathogens can pass between people via a **vector**. This is usually an animal such as a mosquito or a fly.
- Plant pathogens can also be carried on insect vectors. The insect burrows into part of an infected plant, picking up spores. The insect then burrows into an uninfected plant, leaving the spores behind.

These diseases are spread by indirect transmission

Disease	Means of transmission
Malaria	Bite from an infected female Anopheles mosquito. Mosquito carries Plasmodium vivax.
Lyme disease	Bite from a tick, usually Ixodes ricinus.
Sleeping sickness	Bite from a tsetse fly (Africa).
Dutch elm diease	Transmitted by a beetle, Scolytus multistriatus.

Worked example

How can we help prevent diseases spreading by direct contact? **(3 marks)**

washing hands

sneezing/coughing into tissues

washing door handles regularly

Spread of disease

Diseases spread more in warm, cramped, over-populated environments, with poor ventilation and sanitation.

Now try this

1 How do spores spread plant diseases? **(2 marks)**
2 Why do diseases spread more in warm, cramped environments? **(2 marks)**

Plant defences against pathogens

Plants have developed different methods to protect themselves from pathogens.

Plant defences

Cellulose cell walls, lined with waterproof lignin, act as a **physical** barrier to pathogens. However, plant tissues also contain chemicals to kill any pathogens that try to cross this barrier.

These defences are **primary** because they are the first line of defence:

- **Stomatal closing** – Stomata are the small pores on the underside of the leaf (look back at page 63 for the role of stomata). These can close when pathogens are detected.

- **Production of chemicals** – Plants produce many chemicals, such as **terpenes**, **alkaloids** and **phenols**.

- **Callose deposition** – This is when the sieve plates in the phloem (as described on page 61) are blocked with a polysaccharide called **callose**. This prevents pathogens moving through the phloem and limits the spread of the pathogen.

- **Tylose formation** – This is when the xylem is blocked from carrying water by a terpene-filled swelling called a tylose. This prevents pathogens moving through the xylem.

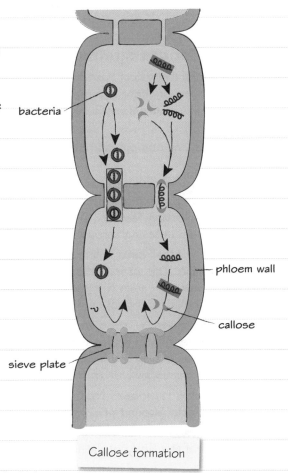

bacteria

phloem wall

callose

sieve plate

Callose formation

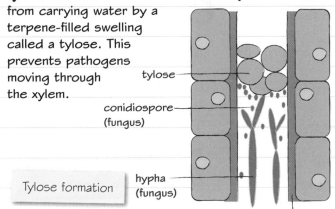

tylose

conidiospore (fungus)

hypha (fungus)

xylem wall

Tylose formation

Now try this

1 Why is stomatal closing a non-specific form of defence? **(1 mark)**

2 Suggest how a plant could prevent an infection by ring rot. **(2 marks)**

Animal defences against pathogens

Animals have developed different methods to protect themselves from pathogens.

Animal defences against pathogens

Skin is an excellent physical barrier against infection from pathogens. Wherever there is a break in the skin, however, the body needs additional defences to prevent pathogens from getting in.

These defences are **non-specific** because they defend against all pathogens:

- **Blood clotting** – When you have a cut, the blood clots around the open wound, using **platelets** and proteins in the blood called **blood-clotting factors**. This forms a scab and prevents pathogens from entering through the open wound.

- **Inflammation** – **Mast cells** in infected tissue release a chemical called **histamine**. This causes blood vessels to widen, or **vasodilate**, and increases the **permeability** of capillaries, so that white blood cells can travel more easily to the site of infection.

- **Mucous membranes** – These line the airways, digestive system, genital areas, ears and nose, and contain **goblet cells** that produce **mucus** to trap any invading pathogens. In the airways, cilia waft the mucus away from the lungs.

- **Stomach acid** – Pathogens that reach the stomach are killed by hydrochloric acid (around pH 2).

- **Expulsive reflexes** – These include coughing, sneezing and vomiting to rid the body of pathogens.

- **Tears and saliva** – These contain lysosome, an enzyme that kills pathogens.

- **Sweating** – Sweat contains chemicals that kill bacteria.

- **Urine flow** – Urinating flushes out pathogens from the bladder area.

- **Commensal bacteria** – These are non-harmful bacteria living on the skin surface that prevent harmful bacteria from growing there.

Inflammation

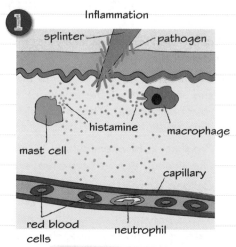

Mast cells release histamine at the site of infection.

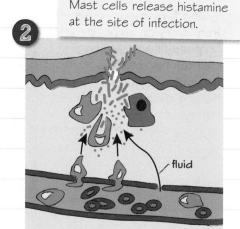

Capillaries become more permeable and allow neutrophils to enter tissue.

What similarities are there between plant and animal primary defences?   **(3 marks)**

Both have a physical barrier.

Both use chemicals.

Both block pathways so that pathogens cannot enter or move around the organism.

Plant cells have cellulose as a physical barrier, animals have skin.

Plants use terpenes, alkaloids and phenols. Animals use histamine.

Plants block the xylem and phloem. Animals block breaks in the skin after injury.

Now try this

1 How might the body defend itself against infection by Salmonella bacteria?   **(2 marks)**

2 Hayfever is caused by the release of histamine when pollen enters the body. What does antihistamine medication do?   **(2 marks)**

Phagocytes

Phagocytes are a type of white blood cell that are part of the non-specific immune defence.

Types of phagocytes

The main two types of phagocytes are **neutrophils** and **monocytes**. Neutrophils are usually found in the blood, although they do also move into the body's tissues.

Monocytes are larger, less abundant and usually found in the tissues, where they develop into **macrophages**.

When the body is infected, chemicals called cytokines are released. Phagocytes are attracted to cytokines and follow the chemical signal to its source in a process called **chemotaxis**.

When a macrophage has engulfed a pathogen, it will display some of the pathogen's antigens on its surface. The macrophage has become an **antigen-presenting cell**.

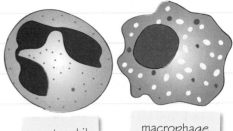

neutrophil   macrophage

Mode of action of phagocytes

1. chemotaxis and adherence of pathogen to phagocyte
2. ingestion of pathogen by phagocyte
3. formation of a phagosome
4. fusion of the phagosome with a lysosome to form a phagolysosome
5. digestion of ingested pathogen by enzymes
6. formation of residual body containing indigestible material
7. discharge of waste materials.

The engulfing of pathogens by phagocytes is made easier by the action of **opsonins**. These are any molecules that enhance phagocytosis. Antibodies that bind to the surface of the pathogen act as opsonins.

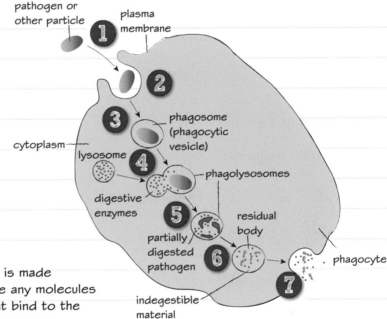

Examination of blood cells

If you look at a blood smear using a microscope, you can identify the neutrophils and red blood cells (erythrocytes).

You will be expected to draw an example of each type of cell, and label the plasma membrane, cytoplasm and nucleus.

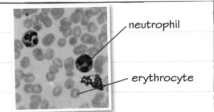

neutrophil

erythrocyte

Worked example

Why do neutrophils have so many lysosomes?
(3 marks)

In phagocytosis, the enzymes in the lysosomes are used to digest the pathogen.

Now try this

Explain the role of antigen-presenting cells. **(3 marks)**

Macrophages also contain many lysosomes.
After digestion, waste materials are expelled by the phagocyte in a process called endocytosis.

Lymphocytes

Lymphocytes are a type of white blood cell involved in the specific immune response.

B lymphocytes

- B lymphocytes are made and mature in the bone marrow. They then migrate to the lymph nodes.
- Each B lymphocyte recognises one specific antigen. When free antigen is detected (**clonal selection**), the B lymphocyte multiplies rapidly (**clonal expansion**) into plasma cells and memory cells. All of these cells are clones.
- **Plasma cells** make **antibodies**, which bind to the antigen, and prevent the pathogen from entering cells. The antibodies make it easier for phagocytes to engulf and destroy the pathogen. The plasma cells then die.
- **Memory cells** stay in the lymph for a long time. They remember the antigen and if it is detected again, the plasma cells against it can be made more quickly, and in greater concentration.

T lymphocytes

- T lymphocytes are made in the bone marrow and mature in the thymus. They also migrate to the lymph nodes.
- Each T lymphocyte recognises one specific antigen, but only when it is displayed on the surface of another cell. When bound antigen is detected, the T lymphocyte starts to multiply rapidly into T killer cells, T helper cells, T regulatory cells and T memory cells.
- **Killer T cells** kill any cells infected with a virus. **Helper T cells** secrete **interleukins**, a type of **cytokine**, which help B lymphocytes to develop. They also stimulate phagocytes to carry out phagocytosis.
- **Regulator T cells** suppress the immune reaction, once the pathogen has been destroyed.

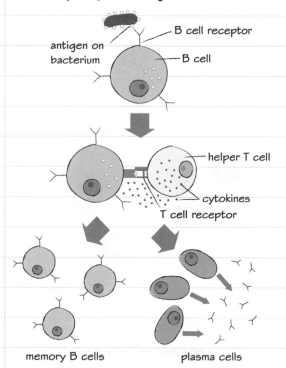

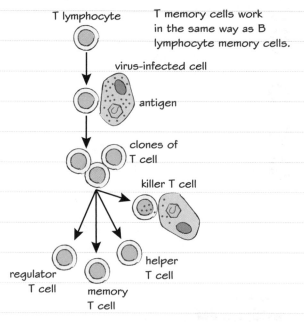

T memory cells work in the same way as B lymphocyte memory cells.

Think back to Module 2, on the secretion of proteins (page 6).

Worked example

Suggest how the plasma cells are different from the B lymphocytes.   **(3 marks)**

increase in rough endoplasmic reticulum / ribosomes

increase in Golgi bodies

increase in mitochondria

Now try this

1 How are the cell organelles involved in the secretion of antibodies from the plasma cells?   **(3 marks)**
2 Explain how killer T cells help to rid the body of a virus.   **(3 marks)**

Immune responses

The primary immune response

The first time a pathogen is detected in the body, the response by the lymphocytes is known as the **primary immune response**.

Antibodies are produced by the plasma cells, but these take a few days to get to the number that will successfully fight off an infection. The antibodies do not stay in the blood, but **T and B memory cells** will remember the pathogen.

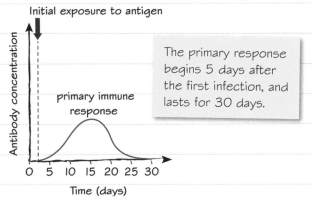

Initial exposure to antigen

The primary response begins 5 days after the first infection, and lasts for 30 days.

The secondary immune response

The second time a pathogen is detected in the body, the response by the lymphocytes is known as the **secondary immune response**.

The plasma cells begin to produce antibodies sooner and at a faster rate. This response also lasts for a longer time. The second response is so rapid you may not even know you have been infected.

Second exposure to antigen

The secondary response begins 1 day after the second infection and lasts for over 30 days.

Antibody structure

Antibodies are Y-shaped molecules made of four polypeptide chains — two heavy chains and two light chains.

The lower part of the antibody is called the **constant region**. This is the same in all antibodies. The top part of the antibody is called the **variable region**. This part of the antibody is different for each antibody, and is **complementary** to the antigen.

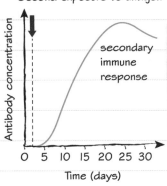

light polypeptide chain

disulfide bridge to hold polypeptides together

hinge region to allow flexibility so molecule can grip more than one antigen

heavy polypeptide chain

variable region

constant region – it may have a site for the easy binding of phagocytic cells

The secondary response will start sooner because the memory cells remember the antigen from the first infection.

The functions of antibodies

Opsonins, agglutinins and anti-toxins are all types of antibody.

Description	Function
Opsonin	Any molecule that enhances phagocytosis by marking an antigen (usually an antibody).
Agglutinin	A substance that causes pathogens to coagulate. One type of agglutinin is a form of antibody that looks like several Y-shaped antibodies attached together.
Anti-toxin	An antibody that binds to toxin, making it harmless.

Antibodies can neutralise pathogens by binding to the pathogen's binding site, preventing it from entering the body's cells.

Worked example

How is the secondary immune response different to the primary response?   **(3 marks)**

More antibodies are produced. Antibodies are produced sooner. Secondary immune response lasts for a longer time.

Now try this

How do agglutinins help the immune response?   **(3 marks)**

Types of immunity

Immunity can either be active or passive, and can be natural or artificial.

Active natural immunity

Active immunity is when there is an active response by the plasma cells to produce antibodies.

If you had an infection, such as chicken pox, your plasma cells would make antibodies for that infection. The second time you encounter the chicken pox virus, you will be immune. This is an example of active natural immunity.

Passive natural immunity

Passive immunity is when antibodies are provided for you. There is no active response from the plasma cells.

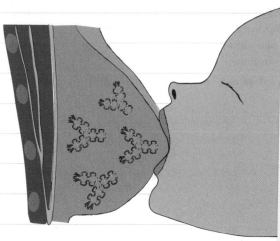

A breastfeeding baby will receive antibodies from its mother in the breast milk. This is an example of passive natural immunity.

Artificial immunity

This is when a person is deliberately exposed to antigens or antibodies from another source.

- Active artificial immunity is when your body is stimulated to produce your own antibodies, by **vaccination** (for more on this topic, see pages 75 and 76). Antigens, such as tetanus, are injected into the body, which causes the plasma cells to produce the specific antibodies.

- Passive artificial immunity is when you are given antibodies from another source. Imagine you've just stood on a rusty old nail in a farmyard. To prevent the possibility of tetanus, you will be injected with antibodies as your treatment.

Autoimmune diseases

An autoimmune disease is when the immune system attacks its own body cells.

Your body cells have antigens on the cell surface that identify your cells as belonging to your body. Usually, this means that the white blood cells do not attack these cells.

When the body stops recognising 'self' antigens it can cause several diseases:

Autoimmune disease	Cause and symptoms
Rheumatoid arthritis	Immune system attacks the body's joints, resulting in stiffness and pain.
Lupus	Immune system attacks cells all over the body. Most common symptoms are fatigue and joint pain.
Type I diabetes	Immune system attacks the cells in the pancreas that make insulin. Insulin is no longer produced and blood sugar levels rise.

Now try this

1. Having a booster vaccination is an example of what type of immunity?   **(2 marks)**
2. Evaluate why active immunity is more useful than passive immunity.   **(3 marks)**

Artificial immunity is often safer than natural immunity. Vaccines contain dead virus particles or just the antigens, and are less likely to cause symptoms than encountering the real pathogen.

The principles of vaccination

Vaccinations are given to people to protect them from being made ill by a pathogen.

How vaccination works

Vaccination is the deliberate exposure of a person to antigens, in order to allow the person to make antibodies to that antigen.

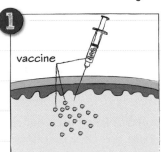

1 Vaccine containing harmless antigenic material from a pathogen is injected into the arm. The immune response starts.

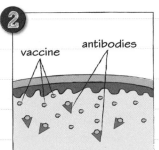

2 B lymphocytes are stimulated to make antibodies and T lymphocytes to kill infected cells.

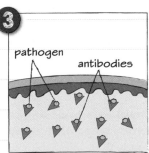

3 If the pathogen enters the body in the future, memory cells will remember the antigen. In the future, the immune response will be rapid and long-lasting. The person is immune to the disease.

How vaccinations are made

The vaccine can be made up of:

- dead pathogen, for example cholera
- the antigens from the surface of the pathogen, for example hepatitis B
- pathogen that has been rendered harmless (attenuated), for example measles
- live microorganisms that are similar but less harmful than the pathogen, for example cowpox to protect against smallpox
- the harmless version of a toxin (toxoid), for example tetanus.

Herd immunity

If most of a population are vaccinated against a particular disease, the population is said to have **herd immunity**. This means that the disease is much less likely to be transmitted through the population. This protects the few individuals who cannot be vaccinated.

Worked example

Describe how a person is vaccinated against a pathogen, such as measles. **(3 marks)**

A vaccine is made containing harmless antigenic material and injected into the person. The B lymphocytes are stimulated to produce antibodies against the antigen. Memory B cells remember the antigen and can produce a fast secondary response if the person is infected with the antigen again.

Vaccines made using dead or attenuated pathogens are less effective than using live pathogen. They often need one or more booster vaccinations later on.

Now try this

1 How do toxoids work? **(3 marks)**
2 Speculate what could happen if vaccination rates for childhood diseases fall below 80%. **(3 marks)**

Vaccination programmes

Vaccination programmes are routinely carried out around the world to protect children from pathogens. For example, you may have been given the polio vaccine as a child.

Vaccination programmes

If a new disease breaks out, and the population are not vaccinated against it, a type of vaccination called **ring vaccination** can be used. All of the people in the area where the disease has broken out are vaccinated to prevent the disease spreading to the rest of the population.

Disease outbreaks worldwide are carefully monitored in order to prevent an **epidemic** (the rapid spread of an infectious disease).

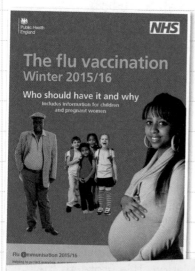

Vaccines are prepared by governments in advance and given to at-risk groups as a preventative measure. For example, flu vaccines are given every year to the elderly, to pregnant women and to young children.

Changes to vaccines

When a new virus appears, a new vaccine is needed to protect against it. A new vaccine might be needed:

- If an animal virus moves into people, for example swine flu.
- If the antigens on a virus mutate and a new vaccine is needed against that antigen, for example influenza.

If most people in a population have been vaccinated against a disease and break outs are very rare then the vaccination program can be stopped, for example smallpox vaccinations worldwide. In 1980, smallpox was declared to have been eradicated.

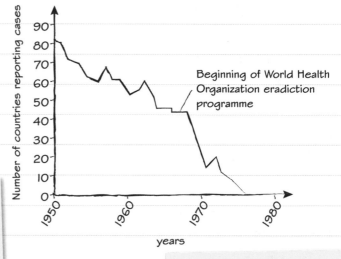

The eradication of smallpox

Imagine there is an outbreak of measles at a school. How can the infection be prevented from spreading through the population?

(3 marks)

Ring vaccination.
Vaccinate everyone in contact with the infected person.
Vaccinate everyone in the area.

Recent measles outbreak

In 2013, there was a measles epidemic in Wales, which resulted in 88 people being hospitalised, and one death.

Vaccination programmes were organised in Welsh schools to help to contain the outbreak.

Now try this

1 What is an epidemic? **(1 mark)**
2 Explain why flu vaccines need to be given every year. **(3 marks)**

Antibiotics

Antibiotics are a group of medicines used to treat bacterial infections.

The benefits of using antibiotics

The first antibiotic to be discovered was penicillin, in 1928. Alexander Fleming discovered that *Penicillium*, a type of fungus, killed bacteria on his Petri dishes. This discovery was later developed into penicillin.

Since the 1920s, many other antibiotics have been discovered, reaching a peak in the 1980s. Some are derived from plants or fungi, while others have been made in the laboratory by modifying already known antibiotics. Thanks to antibiotics, we can now easily treat infected wounds or bacterial infections, such as tuberculosis. However, fewer new antibiotics have been discovered in the last two decades. This is a problem because of the worldwide increase in antibiotic resistance in bacteria.

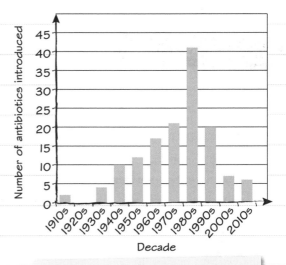

The number of antibiotics introduced between 1910 and 2010

Antibiotic resistance

Since the first antibiotics were used, some of the bacteria that they killed have become **resistant** to them. This means that the antibiotic no longer kills them. Some infections have to be treated with a combination of different antibiotics to get around this problem.

MRSA (methicillin-resistant *Staphylococcus aureus*) and *Clostridium difficile* are two types of bacteria that are resistant to most antibiotics and so are very difficult to treat. Outbreaks of these bacteria in hospitals can shut down whole wards; the wards need to be deep cleaned to get rid of the infection.

Sources of new medicines

1 **Nature** — New medicines are constantly being searched for. Fungi and plants have been good sources of new medicines in the past, for example, aspirin was found in the bark of the willow tree. Current searches are taking place in the rainforests.

2 **Personalised medicine** — It has been known for some time that not all medicines work for all people. This is likely to be due to genetic differences between people. In the future it is possible that the medicine you are given will be dependent on your genetic profile.

3 **Synthetic biology** — This is the construction of new biological parts or cells. For example, developing synthetic bacteria that deliver chemotherapy drugs to cancer cells.

Worked example

New antibiotics are being made all the time, but it is expensive and labour intensive.

Why are new antibiotics needed?   **(2 marks)**

Bacteria become resistant to current antibiotics.
Some bacteria / MRSA / C. *difficile* are resistant to most antibiotics.

Careful management of antibiotics is needed to control the rate at which bacteria become resistant to them.

Now try this

1 Why is conservation important in the search for new medicines?   **(2 marks)**
2 Explain why human behaviours are leading bacteria to become resistant to antibiotics.   **(2 marks)**

Exam skills

This question is about pathogens and immunity, and uses knowledge and skills that you have already revised. Look at pages 67–77 to refresh your memory.

Worked example

(a) What type of pathogen causes influenza? **(1 mark)**

virus

(b) Name two ways that influenza can spread. **(2 marks)**

coughing / sneezing
touching infected surfaces

> Influenza is spread in droplets of saliva and mucus that are sneezed or coughed out of an infected person. If an infected person sneezes onto a surface, or onto their hands and then touches a surface, then touching that surface can also spread influenza.
>
> Disposing of tissues, hand washing and cleaning surfaces such as door handles can all reduce the spread of influenza.

(c) What primary defences does the body have to protect against infection from influenza in air-borne droplets? **(2 marks)**

cilia wafting mucus up to the mouth

stomach acid / hydrochloric acid in stomach

> There are many other primary defences, including blood clotting, inflammation, expulsive reflexes, tears, sweat, urine flow and commensal bacteria.

(d) Explain how B lymphocytes would respond to influenza if the body had never encountered that strain of influenza before. **(4 marks)**

The B lymphocyte that is specific to the influenza antigen differentiates into plasma and memory cells.

The plasma cells produce antibodies that are complementary to the influenza antigen.

The memory cells remember the antigen.

> There are many B lymphocytes in the lymph nodes, but only one has antibodies on the surface that are a complementary shape to the influenza antigen.

(e) Draw the shape of the graph showing the concentration of antibodies if the body were to detect the same antigen again. **(2 marks)**

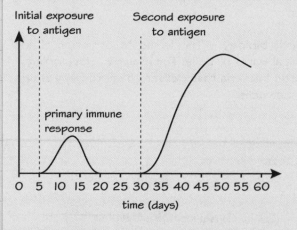

> The graph line should be steeper and drawn higher than the primary immune response.

(f) What name is given to the response of the B lymphocytes to the detection of an already known antigen? **(1 mark)**

secondary immune response

(g) What type of immunity is being shown by the graph above? **(2 marks)**

active, natural immunity

> This type of immunity is natural because the person is being exposed to a pathogen in a natural way. It is active, because the body's B lymphocytes are being stimulated to produce antibodies.

Measuring biodiversity

Biodiversity is a measure of the number and variety of species living in an ecosystem.

What is biodiversity?

Biodiversity can be considered at several different levels:

- Habitat biodiversity – the number of different habitats in an ecosystem.
- Species biodiversity – examining the different species living in a habitat and measuring the species richness and species evenness. This can be measured using the Simpson's index of diversity (for more detail, see page 81).
- Genetic biodiversity – genetic variation between individuals of the same species.

> **Species richness** – the number of different species in a habitat.
>
> **Species evenness** – the number of individuals of the same species in a habitat.

The importance of sampling

Sampling is selecting a small area or areas in a habitat to study. It is better to sample a habitat because studying every part of it would be too time-consuming.

Random sampling is selecting areas of the habitat to study in a random manner. This eliminates **bias**. Enough samples have to be taken so that the data are **representative** of the habitat.

Non-random sampling can be:

- Opportunistic – deliberate samples chosen during data collection in order to get representative data.
- Stratified – choosing two different areas in a habitat to sample.
- Systematic – samples taken at fixed intervals across a habitat, e.g. transect.

Advantages and disadvantages of random and non-random sampling

Random sampling is achieved by using randomly selected numbers as coordinates. Only those coordinates are sampled, using quadrats or a method of animal collection.

In non-random sampling, a decision is made about which areas to sample. Each of these methods has their own advantages and disadvantages.

	Advantages	Disadvantages
Random sampling	Data are not biased by selective sampling.	Rare species may be missed, leading to an underestimate of biodiversity.
Non-random sampling	Ensures all areas of a habitat are sampled. Can identify changes in a habitat.	Data may be biased and some areas may be over-represented.

Worked example

What are the important features of sampling?   **(3 marks)**

Should not be too time-consuming.

Should not be biased.

Should be representative of the species in the habitat.

You should understand that counting every single plant or animal in a habitat is far too time-consuming.

Your sampling method should include all of the species that it is possible to measure in a habitat, including rare or small species.

Now try this

1 Which is more important for biodiversity, species richness or species evenness?   **(3 marks)**
2 Compare random and non-random sampling.   **(4 marks)**

Sampling methods

Practical skills There are a number of different sampling methods that can be used, depending on the type of organism that is being sampled.

Quadrats and transects

Quadrats and transects are methods of sampling plants.

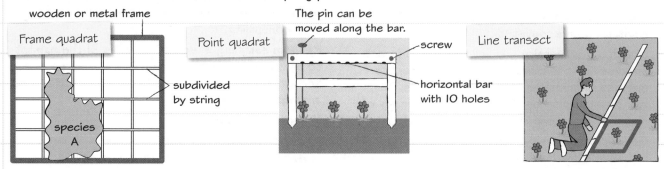

wooden or metal frame

Frame quadrat

subdivided by string

species A

The pin can be moved along the bar.

Point quadrat

screw

horizontal bar with 10 holes

Line transect

In a **frame quadrat**, the species richness and species evenness are measured. Each square is 4% of the quadrat area. Since species A occupies about five squares, it covers 20% of the ground.

In a **point quadrat**, the species touching each pin are counted. Count all the plants touched by the pin. Two or more species may cover the same ground where plants overlap.

Transects are used to measure the change in species across an area, for example a sand dune.

- With a **line transect**, quadrats are placed at regular intervals along the line.
- With a **belt transect**, the species richness and species evenness are measured between two lines.

Collection methods

These methods are used to collect animals within quadrats.

Pitfall trap

slate

supporting stone

ground level

jam jar

soil

bait to attract animal being sampled

animals sucked in here

suck

Pooter

gauze over sucking tube

- **Pitfall traps** are set below the ground so that small invertebrates will fall in.
- Small invertebrates are pulled into **pooters** on a straw.
- **Kick sampling** collects animals that live at the bottom of a stream or pond. A net is held under the water while the streambed is agitated by kicking for several minutes.
- A **sweep net** collects animals that live in a field or long grass. The net is swept through the field to collect insects.

In all cases, the animals are released unharmed after counting.

Worked example

When would you use a line transect?   **(2 marks)**

In a large habitat, where the habitat changes along the line, for example along sand dunes.

Now try this

A farmer wants to know if a field needs spraying with pesticides. Which collection method should be used on the field and why?   **(3 marks)**

Simpson's index of diversity

Measuring biodiversity

The Simpson's index of diversity is a measure of the level of biodiversity of the habitat.

In order to calculate Simpson's index, you need to measure the species richness and species evenness (as defined on page 79) within a habitat.

This can be done by using quadrats or a transect, and counting the number of different species and number of individuals of each species within each quadrat.

Maths skills **Interpreting Simpson's index**

This is the formula for Simpson's index:

This index gives a number between 1 and 0.

n is the number of individuals of each species.

$$D = 1 - \left[\Sigma \left(\frac{n}{N} \right)^2 \right]$$

N is the total number of individuals in a habitat.

Σ means 'sum of'.

The closer the score is to 1, the higher the level of biodiversity.

Maths skills **Using Simpson's index of diversity**

Here are some data collected from quadrats in two different habitats:

Species	Area 1 Number of individuals	n/N	$(n/N)^2$	Area 1 Number of individuals	n/N	$(n/N)^2$
Dandelion	3	0.3	0.09	6	0.6	0.36
Daisy	2	0.2	0.04	0	0.0	0.00
Buttercup	1	0.1	0.01	0	0.0	0.00
Clover	3	0.3	0.09	4	0.4	0.16
Thistle	1	0.1	0.01	0	0.0	0.00
	$N = 10$	$\Sigma =$	0.24	$N = 10$	$\Sigma =$	0.52
		$D = 1 -$	0.24		$D = 1 -$	0.52
		$D =$	**0.76**		$D =$	**0.48**

Both areas have an N of 10, but the Simpson's index for area 1 is 0.76 and is 0.48 for area 2. Therefore, area 1 has a higher level of biodiversity than area 2. It is the number of different species as well as the number of individuals of each species in each habitat that is important for biodiversity.

Worked example

Explain which area in the table above has the highest species evenness. **(3 marks)**

Species evenness is the number of individuals of that species in a habitat. Area 2 has the highest species evenness, because there are 6 dandelions.

Learn the definitions for species richness and species evenness.

Area 1 has lower species evenness, but higher species richness.

Now try this

1 Suggest why a habitat with low biodiversity is less stable. **(3 marks)**

2 Calculate the change in biodiversity for area 2 if there are 2 dandelions, 2 buttercups, 1 daisy, 2 clovers and 1 thistle. **(3 marks)**

Factors affecting biodiversity

Genetic diversity

Genetic diversity is the amount of genetic variation within a population. Usually a population with a large number of individuals has a high genetic diversity.

Populations with a small number of individuals, such as in a rare species, will have low genetic diversity.

We can observe the level of genetic diversity by looking at the **phenotypes** (observable traits) of the population. Many differences in phenotype, e.g. fur colours, will indicate high genetic diversity.

High genetic variation

Maths skills — Calculating genetic diversity

Genetic diversity is caused by the number of **alleles** of each gene. If a gene has two or more alleles it is called **polymorphic**.

At each **locus** (position on the chromosome), there are two alleles. If the alleles are different they are **heterozygous**. If the alleles are the same they are **homozygous**.

Genetic diversity can be calculated by working out the number of genes that are polymorphic have heterozygous alleles, as a proportion of all genes:

$$\text{proportion of polymorphic gene loci} = \frac{\text{number of polymorphic gene loci}}{\text{total number of loci}}$$

Human influences

Many human activities may affect the level of biodiversity in a habitat:

Factor	How it affects biodiversity	Example
Human population growth	The increased human need for land, food and resources reduces the habitats and food sources for other species.	Cutting down forest to use land for housing.
Monoculture	When only one species of a plant is planted in an area at a time, there is less space for other species.	Planting a forest of palm trees for palm oil.
Climate change	Increased human activity increases climate change, which affects the habitats of other species.	Increased flooding of low lying areas.
Extinction	Increased human activity increases the extinction rate of many species.	The Japanese river otter became extinct in 2012 due to hunting and human pollution of its habitat.

Worked example

Explain why a single dog pedigree would have a low genetic diversity. **(3 marks)**

Genetically isolated population because breeding is controlled by humans. This leads to a lower proportion of polymorphic gene loci.

A lower proportion of polymorphic gene loci means that many individuals in the breeding population share the same alleles.

Populations with low genetic diversity are more likely to carry recessive disease alleles. For example, many dog breeds are more prone to certain genetic conditions.

Now try this

1 If an individual has 7500 polymorphic gene loci out of 30 000 gene loci, what is the proportion of polymorphic gene loci in that individual? **(2 marks)**

2 Comment on the ways that climate change can affect habitats. **(4 marks)**

Maintaining biodiversity

Why maintain biodiversity?

There are several reasons for maintaining biodiversity:

- **ecological** – protecting keystone species and maintaining genetic variation.
- **economic** – services and products that plants and animals provide for us, such as maintaining the nutrients in the soil, timber, fruit and honey.
- **aesthetic** – protecting landscapes: enjoying the natural beauty of the planet, for example going on holiday or sitting in a garden.

In situ methods of conservation

In situ means that the plants or animals are being kept in their natural environment.

Conservation zones can be created for marine wildlife. In these areas, no fishing or hunting is allowed.

Terrestrial animals can live inside **wildlife reserves**. In these areas, no hunting is allowed and rangers protect the animals.

Advantages	Disadvantages
Biodiversity in that ecosystem is protected.	Difficult to carry out breeding programmes Animals may not have access to medical care.

Ex situ methods of conservation

Ex situ means that the plants or animals are being kept outside of their natural environment.

Seeds of plants can be kept in a **seed bank** to protect them for future generations. Plants may be grown in **botanical gardens**, such as the gardens at Kew (West London), so that the plants can be conserved and studied.

Animals may be kept in a **zoo** or **aquarium**.

Advantages	Disadvantages
• Can carry out breeding programmes • Prevent extinction of species • Protection from hunting • Animals will have access to medical care	• Does not protect the biodiversity of the original habitat • Will lose some genetic diversity due to fewer breeding pairs • Difficult to reintroduce species back into the wild

Conservation agreements

- The Convention of International Trade in Endangered Species (CITES), 1973: signed by most countries, this prevents the trade of endangered wild plants and animals and animal parts, e.g. ivory.
- The Convention on Biological Diversity, Rio Earth Summit, 1992: signed by more than 150 countries. It is designed to promote worldwide cooperation towards sustainable development and increasing biodiversity.
- The Countryside Stewardship Scheme (CSS), UK: aims to increase biodiversity across the UK.

Worked example

Explain what type of conservation is best for conserving a critically endangered species. **(3 marks)**

Ex situ conservation.

Species would be protected from hunting.

Breeding programmes would increase numbers.

Worldwide breeding programmes of rare species aim to increase the number in the population, before releasing back into the wild in a conservation area.

Modern techniques, such as freezing sperm/eggs and IVF, could also be used to increase population numbers.

Now try this

1 Seeds can be kept in a seed bank. What are the disadvantages of this with regard to genetic variation? **(3 marks)**
2 Explain the advantages and disadvantages of treaties such as CITES. **(4 marks)**

Classification

Classification is the organisation of organisms into groups based on their characteristics.

Levels of classification

Organisms can be classified by their physical characteristics. This is known as **taxonomy**.

Organisms with similar physical features are grouped together into gradually smaller categories.

Traditionally, all organisms are in one of five kingdoms: Animalia, Plantae, Fungi, Protoctista, and Prokaryotae. However, more recently, a higher taxonomic rank has been added, called the domain.

There are three domains: Archaea, Eubacteria and Eukaryotae (or Eukarya).

Taxonomic hierarchy for lions

Domain	Eukarya
Kingdom	Animalia
Phylum	Chordata
Class	Mammalia
Order	Carnivora
Family	Felidae
Genus	Panthera
Species	leo

Phylogenetic trees

Organisms can also be classified by their genetic and molecular make up. This is called **phylogeny**. Organisms that are similar are closer together on a phylogenetic tree.

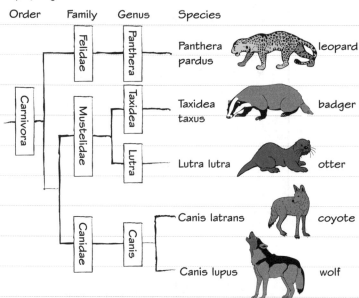

Binomial classification

All species are given two Latin names. For example, lions have the binomial name *Panthera leo*. The first name is the genus and the last name is the species. Species that are very similar will be part of the same genus. For example, leopards are *Panthera pardus*.

The main advantage of the binomial system is that it gives species a universal name. All scientists worldwide use the same name for each species.

Wolves are more closely related to coyotes than leopards, so they are closer together on the phylogenetic tree.

Worked example

Suggest which criteria a taxonomist would take into account when classifying a new species. **(4 marks)**

anatomy / morphology

biochemistry

genetics

behaviour

Now try this

1 In the above phylogenetic tree:
   (a) Which genus do the coyote and the leopard belong to? **(1 mark)**
   (b) Which family do the badger and the otter belong to? **(1 mark)**
2 What does taxonomic hierarchy mean? **(2 marks)**

The five kingdoms

Characteristics of the five kingdoms

All organisms were originally placed into one of the five kingdoms according to observations of their features. These days, we also look at characteristics of their cell structure (look back at page 5), anatomy and form of nutrition. **Autotrophic** organisms make their own food, usually using energy from the sun, whereas **heterotrophic** organisms depend on other organisms for nutrition.

Kingdom	Characteristics
Prokaryotae	single-celled, prokaryote, heterotrophic, cell wall (usually made of peptidoglycan)
Protoctista	single-celled or simple body form, eukaryote, heterotrophic and autotrophic, no cell wall
Fungi	single-celled or multicellular, eukaryote, heterotrophic, chitin cell wall, cells are multinucleate, has hyphae
Plantae	multicellular, eukaryote, autotrophic, cellulose cell wall
Animalia	multicellular, eukaryote, heterotrophic, no cell wall

Domain system

Organisms have, more recently, also been divided into **domains** in order to clarify the relationships between them.

There are three domains: Archaea, Eubacteria and Eukaryotae. The Prokaryotae kingdom divides between the Archaea and Eubacteria domains, and the other four kingdoms all belong to the Eukaryotae domain.

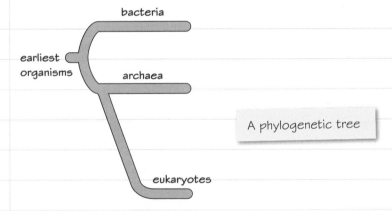

A phylogenetic tree

Classification and phylogeny

Originally, the classification of species used anatomy and behaviour to classify species into **taxa** (groups).

However, modern techniques, such as DNA sequencing and examining the RNA sequence, can also be used. This shows how closely species are related to each other and is called **phylogeny**. These modern techniques support the domain classification of species.

Phylogeny does not attempt to put organisms into groups; rather it places each species along a continuum in a **phylogenetic tree** (see opposite). These trees show the evolutionary relationship between organisms.

Worked example

Compare and contrast the domain and kingdom systems of classification. **(3 marks)**

Both systems look at features of the cells to classify species.

Both systems use modern techniques to compare DNA and RNA sequences between species.

Kingdoms are a sub-division of domains.

The Eubacteria domain and the other two domains separated along different evolutionary lines. The Archaea domain separated from the Eukaryotae domain later.

Now try this

1 A new species has been discovered that has no cell walls and is autotrophic.
   Explain which kingdom it belongs to.   **(2 marks)**

2 How is classification different to phylogeny?
   (4 marks)

3 How is DNA used to construct phylogenetic trees?   **(3 marks)**

Types of variation

Variation is differences between individuals. Variation is caused by either genetics or the environment or both.

Causes

Some characteristics, such as eye colour, are due to genetics. Other traits, such as a scar, are caused by something that happened in the environment. Many characteristics have a genetic and environmental influence, such as height and weight.

- **Interspecific variation:** between individuals of different species.
- **Intraspecific variation:** between individuals of the same species.

Continuous and discontinuous variation

Within a species, there is also continuous and discontinuous variation.

Discontinuous variation: a characteristic can be placed into a discrete category, such as blood group.

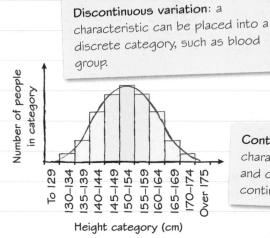

Continuous variation: characteristics have a range and can be placed on a continuum, for example height.

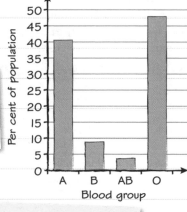

🖩 **Maths skills** **How to use Student's *t*-test to compare two sets of data**

The Student's *t*-test is used to determine if two sets of data are significantly different from each other. First, a null hypothesis is written that states that there is no difference between the two data sets. Then, the two sets of data are subjected to the formula:

$$t = \frac{\bar{X}_1 - \bar{X}_2}{\sqrt{\dfrac{S_1^2}{n_1} + \dfrac{S_2^2}{n_2}}}$$

— $\bar{X}_1$ and $\bar{X}_2$ are the means of the first and second set of data.

— S_1 and S_2 are the standard deviations of the first and second set of data.

— n_2 is the total of the values in the second data set.

— n_1 is the total of the values in the first data set.

Worked example

Two sets of plants have been grown, one with and one without fertiliser. Is the height of the plants grown with fertiliser significantly different? **(4 marks)**

						n	$\bar{X}$	S	S^2	S^2/n
Height of plant without fertiliser / cm	10	6	8	12	9	45	2	2.2	4.84	0.108
Height of plant with fertiliser / cm	11	12	14	9	16	62	12.4	2.7	7.29	0.118
			$X_1 - X_2 = -3.4$							$\Sigma = 0.225$
										$\sqrt{} = 0.48$
										$t = 7.08$

The *t*-test score is -7.08. This is now compared to a Student's *t*-test table (the minus sign makes no difference). If the number is larger than the critical value at $p = 0.05$, then the results are significant and we must reject the null hypothesis.

In this example, there are 10 pieces of data, so the degree of freedom is 9. The critical value for $p = 0.05$ is 2.262. The *t*-test score is greater than 2.262; therefore there is a less than 5% probability that these results happened by chance and we must reject the null hypothesis.

Now try this

Suggest what the results would show if the *t*-test score in the above example was 1.12. **(3 marks)**

Evolution by natural selection

Adaptations of organisms to their environment

All organisms have anatomical, physiological and behavioural adaptations to their environments that help them to survive. For example, animals that live in the extreme cold:

- have thick fur (anatomical)
- will build up thick layers of blubber under the skin in winter (physiological)
- huddle together for warmth (behavioural).

Species that live in similar environments tend to have similar anatomical features, even though they belong to different taxonomic groups. For example, the marsupial mole belongs to an order of mammals called *Notoryctemorphia*, and the placental mole belongs to an order of mammals called *Soricomorpha*. However, these two species look very similar, despite having a very different evolutionary past.

Placental mole

Marsupial mole

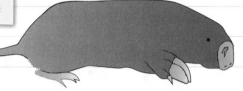

The mechanism of natural selection

1. All organisms must compete to survive and overcome **selective pressures**. Some examples of selective pressures are food, water, predation and disease.

2. All organisms show **variation**.

3. **Selection**: some individuals will be better adapted to overcoming a particular selective pressure than others and are more likely to survive long enough to have offspring.

4. **Adaptation**: the offspring will have the same advantageous characteristics as the parents if the characteristic is the result of genetic information. Over time, more of a population will have the advantageous characteristic.

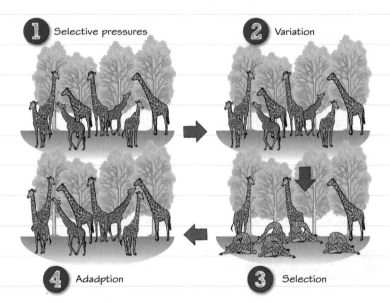

1 Selective pressures   2 Variation

4 Adadption   3 Selection

Worked example

Why do species from similar environments have similar physical traits? **(2 marks)**

same selective pressures

same adaptations

Particular selective pressures will drive evolution in a particular direction. For example, both species of moles need to be adapted to dig and both have evolved long front claws.

Now try this

1 Suggest the adaptations an animal living in a desert environment might have. **(3 marks)**

2 Predict what would happen to a predated species living in a green environment, where some individuals were blue and some were green. **(4 marks)**

Evidence for evolution

Evolution is the theory that the differences between modern living organisms are because of changes that happened under the influence of natural selection over a very long period of time.

Darwin and Wallace

In the mid-1800s, Charles Darwin and Alfred Russel Wallace independently came up with a theory of evolution by natural selection. Natural selection is a mechanism to explain how evolution could have occurred. The two men collaborated and co-published a paper in 1858. Darwin published his book, 'On the Origin of Species' in 1859.

Evolution of resistance

Evolution in some species has implications for humans. This can be seen in resistance to chemicals. As we invent new pesticides to kill pests on crops, some of the pests can develop **pesticide resistance**. This means that there will be less **yield** from the crop.

The same thing happens with bacteria and antibiotics, as certain bacteria develop **antibiotic resistance**. Look back at page 77 for a reminder.

Evidence for evolution

There is an overwhelming amount of evidence for evolution, from different sources.

- The fossil record – we can examine **extinct** species and compare ancient organisms with modern ones. The age of the fossils can be measured by **dating** the rocks that they are found in. The fossil record gives us the fossils in **chronological** order.

Some species, such as the horse, have an entire fossil record, showing all of the **intermediate** forms.

Eohippus   Oligohippus   Merychippus   Pliohippus   Modern horse

- DNA – the DNA sequences of species can be compared to see how closely related species are.

Species A - G C C A T A A C C T G A G G -
Species B - G C C A T A T A C T G A G G -
Species C - G C C A C A T A G T G A G G -
Species D - G C C A C A T A G T A A G G -
                ↑   ↑↑↑   ↑

The arrows show the bases that are different across species.

- Molecular evidence – the amino acid sequence of common proteins, such as cytochrome c, can be compared to see how closely related species are. Cytochrome c is an essential protein involved in respiration and is common to almost all organisms.

Worked example

Suggest what could happen in the future if no new antibiotics are found.   **(4 marks)**

increased antibiotic resistance

infections fail to be treated, even with multiple antibiotics

increased death from bacterial infection

increased use of older techniques / salt bathing / maggots or modern techniques such as bacteriophages / disinfectants / antiseptics

Now try this

1 From the DNA sequences above, explain which two species are the most closely related.   **(2 marks)**
2 Explain the problems with relying on the fossil record for evidence of evolution.   **(3 marks)**

Exam skills

This question is about classification and variation. Look at pages 84–88 to revise classification, the five kingdoms, types of variation and evolution.

Worked example

(a) Put the taxa in the correct order:
Domain, Kingdom, Species, Class,
Phylum, Family, Genus, Order   **(2 marks)**

Domain Kingdom, Phylum, Class, Order, Family, Genus, Species.

> Think of a mnemonic to remember the order of the taxa, for example: **D**o **K**oalas **P**refer **C**hocolate **O**r **F**ruit **G**enerally **S**peaking?

(b) Why is it useful for all species to have a binomial name?   **(1 mark)**

All scientists worldwide will call the species by the same name.

> Before binomial classification, species often had long, descriptive names, plus a common name. These names were often different in different countries, leading to a lot of confusion!

(c) (i) On the phylogenetic tree, which two species are the most closely related?   **(1 mark)**

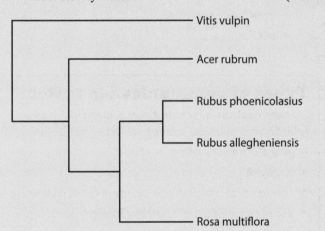

- Vitis vulpin
- Acer rubrum
- Rubus phoenicolasius
- Rubus allegheniensis
- Rosa multiflora

> Phylogenetic trees are based on comparisons of DNA, RNA, and amino acid sequence in key proteins, as well as similarities in physical characteristics.

Rubus phoenicolasius and Rubus alegheniensis.

(c) (ii) How do you know?   **(1 mark)**

They have the same genus.

> Species in the same genus are more similar than species from different genera.

(d) What are the features of all species in the Plantae kingdom?   **(2 marks)**

All species in the Plantae kingdom are autotrophic. Their cells are eukaryotic and contain chloroplasts.

> Autotrophic means that they make organic compounds from inorganic compounds, using light or chemical energy.
> Eukaryotic means that the cells contain a nucleus, as well as other organelles.

(e) Plants in a species can have a variety of heights. What type of variation is this?   **(1 mark)**

Continuous variation.

> Continuous variation has a range of values, e.g. weight or height.

(f) In a shaded environment, taller plants have a selective advantage. Explain how natural selection leads to taller plants.   **(4 marks)**

The amount of sunlight is a selective pressure. Taller plants are more likely to get enough sunlight and survive. Taller plants are more likely to cross-pollinate with other plants. Seeds from taller plants will share the same tall characteristics.

> A selective pressure is anything that can affect the survival of an organism. For example, competition for resources, predation or disease.

The need for communication

Animals and plants need communication systems so they can respond to internal and external environmental changes and coordinate activities between different organs.

The need for a constant internal environment

Animals and plants depend upon enzyme-controlled reactions inside cells to complete key life processes, such as respiration. Certain conditions affect the way enzymes function and can affect their efficiency:

- ✓ temperature
- ✓ pH
- ✓ water balance (water is the medium for reactions)
- ✓ freedom from toxins
- ✓ excess inhibitors.

Responding to change

Changes in the external environment could lead to internal changes if an organism fails to respond. Key stimuli to consider include:

- light
- temperature
- soil pH, mineral content, water content
- salinity (for water-dwelling organisms)
- predators/prey ('fight or flight' response).

In animals, the internal environment is formed by tissue fluid. Cellular processes can change the following:

- temperature
- carbon dioxide concentration
- urea concentration
- pH.

The role of communication systems

Communication systems enable multicellular organisms to detect and respond to changes in their environment.

Cells that detect external or internal changes are often in different parts of the body to the cells that bring about the responses, hence the need for effective communication systems.

Types of communication system

Communication systems have a chemical basis (plants and animals) and/or nervous basis (animals).

A good communication system:

- is rapid
- can lead to short-lived or long-term responses
- enables cell-to-cell communication
- can be specific
- covers the whole organism.

The two communication systems in animals

Nervous system: interconnected network of neurones which signal to each other via electrical impulses and synapses.

Hormonal system: Endocrine cells release chemicals into the blood stream which are subsequently transported to target cells.

Worked example

Explain why it is important that our bodies regulate temperature, pH and inhibitor levels. **(3 marks)**

High temperatures and extremes of pH can cause enzymes to **denature**. Inhibitors can be **competitive** or **non-competitive** and will also affect the **rate** of enzyme action.

This question is really about enzyme action. Think back to what you learnt in Module 2 about enzymes and be sure to include key terminology in your answers.

Now try this

1 Explain the need for communication systems in multicellular organisms. **(3 marks)**
2 Give three internal stimuli which must be monitored and maintained at a steady state. **(3 marks)**

Principles of homeostasis

Homeostasis is the maintenance of a constant internal environment. You need to understand how certain conditions in the human body are maintained in homeostasis.

Homeostatic processes in the human body

Condition	Receptor	Effector	Nervous or hormonal control
Body temperature	Peripheral temperature receptors	Skin, lungs, liver, skeletal muscles	Nervous
Blood glucose concentration	Alpha and Beta cells in the pancreas	Liver and skeletal muscle	Hormonal
Water potential of the blood	Hypothalamus	Kidney (collecting ducts of nephrons)	Hormonal
Carbon dioxide concentration (in blood)	Chemoreceptors	Heart, lungs	Nervous

- **Receptors** detect changes in the **condition**.
- **Effectors** make further changes as a result.

The nature of the change depends on whether the feedback is **positive** or **negative**.

Enzymes

All the conditions in the table opposite affect the way enzymes work. Some enzymes are very sensitive to changes in conditions. Increases in temperature and significant pH changes can cause enzymes to denature, which means they cease to perform their function properly.

Cell signalling

Hormones allow cells to communicate between each other over long distances. Nervous impulses allow communication between adjacent cells.

Negative feedback

Negative feedback is an essential part of homeostasis. Effectors **reverse** changes in the condition to restore optimum levels.

negative feedback

blood glucose
optimum
concentration

The control of blood glucose concentration by insulin is an example of negative feedback. When the blood glucose concentration rises, receptors sense a change and the pancreas secretes insulin to reduce it.

Positive feedback

Positive feedback is **not** part of homeostasis. Effectors **reinforce** change which causes the condition to deviate further from optimum levels.

positive feedback

original oxytocin
concentration

During childbirth, as the cervix begins to stretch, oxytocin is released. This causes uterine contractions, which stretch the cervix more. This is an example of positive feedback.

Worked example

Alpha cells detect a decrease in the blood glucose concentration. They release glucagon. Glucagon causes liver cells to break down glycogen and release glucose into the blood stream. The concentration of glucose in the blood increases. Alpha cells stop secreting glucagon.

Identify the following in the above paragraph:
- stimulus – low blood glucose concentration
- receptor – alpha cell
- effector – liver cells   **(3 marks)**

Now try this

1. What type of feedback does homeostasis rely on?
   (1 mark)

2. If you failed to respond to an increase in body temperature, what would happen and why?
   (3 marks)

3. Explain how the skin could be viewed as a 'receptor' and an 'effector'.   **(2 marks)**

Use the terms stimulus, receptor and effector when discussing all homeostatic pathways.

Temperature control in endotherms

Endotherms, such as birds and mammals, are organisms that can regulate their own body temperature.

Temperature and reaction rates

- Body temperature plays a key role in metabolism.
- Temperature directly affects the rate of reactions – an increase of 10°C doubles the reaction rate.
- Successful collisions between substrates and enzyme active sites result in reactions occurring.
- High temperatures cause enzymes to denature. Remember that low temperatures do not denature enzymes; they simply slow reaction rates.

The physiological response

Vasoconstriction: reduces blood flow to the skin through the contraction of muscles in the walls of the arterioles.

Vasodilation: increases blood flow to the skin through the relaxation of muscles in the walls of the arterioles.

Due to their adaptations, endotherms can live successfully in a wider range of environments than ectotherms.

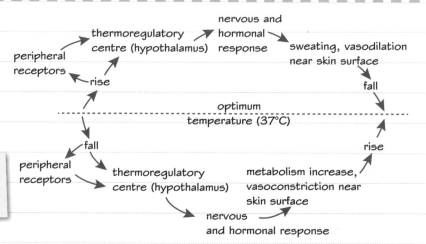

Component of the body involved	Core body temperature too high	Core body temperature too low
Sweat glands in skin	More sweat, more heat loss through evaporation	Less sweat, less heat loss through evaporation
Hairs on skin	Hairs lie flat	Hairs rise, trapping a layer of insulating air
Liver cells	Reduced rate of metabolism	Increased rate of metabolism, more heat from exergonic reactions
Skeletal muscle cells	No spontaneous contraction	Spontaneous contraction (shivering)

The behavioural response

- Endotherms have to eat more than cold-blooded animals (ectotherms) to maintain their body temperature; they also hibernate during cold periods.
- At high temperatures they breathe more (as seen in dogs) and shelter from the sun.

Advantages and disadvantages of being an endotherm

Advantages	Disadvantages
Constant body temperature	High demand for food
Active even when cool	Energy required to
Can live in cold environments	maintain temperature

Worked example

A student was asked what might happen in the body to maintain core body temperature if they went outside on a cold, winter's day. Here is the student's response: "When the outside temperature is too cold, capillaries in the skin dilate to heat the skin up."
Identify the errors in the above sentence. **(2 marks)**

Arterioles not capillaries are involved in this response. They would actually constrict not dilate. The skin would not 'heat up' as a result.

Now try this

1 Why do endotherms need peripheral temperature receptors? **(2 marks)**
2 Why do penguins huddle together? **(2 marks)**

Temperature control in ectotherms

Ectotherms, such as reptiles, fish and insects, cannot regulate their own body temperature and have to alter their behaviour to maintain their body temperature within an effective range.

The behavioural response

To warm up:

☑ expose body to the sun

☑ turn to face sun to maximise the surface area exposed

☑ increase breathing

☑ hair rise and shivering

To cool down:

☑ hide from the sun

☑ turn away from the sun to minimise the surface area exposed

☑ decrease exposed surface area

☑ increase breathing.

Advantages and disadvantages of being an ectotherm

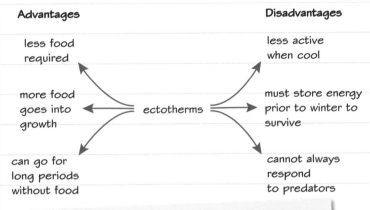

Advantages		Disadvantages
less food required		less active when cool
more food goes into growth	ectotherms	must store energy prior to winter to survive
can go for long periods without food		cannot always respond to predators

Ectotherms can go for long periods without food and some reptiles can live to over a hundred years old (e.g. giant tortoise).

Worked example

 Practical skills

Six human subjects (endotherms) were given 500 ml of iced water to drink. Their temperatures were taken immediately before the drink, then at 5-minute intervals for 20 minutes afterwards.
The results are shown in the table.

		Temperature (degrees centigrade)				
			Minutes after drink			
Subject	Sex	Before	5	10	15	20
1	m	36.4	36.4	36.8	36.6	36.4
2	m	36.8	36.6	37.2	36.8	36.8
3	f	36.9	36.7	36.9	36.9	36.9
4	m	36.8	36.8	37.2	37.1	37.0
5	f	37	37.0	37.2	37.1	37.0
6	f	37.1	37.2	37.4	37.3	37.1

(a) Identify an error in the table. **(1 mark)**

The initial temperature of Subject 5 is only recorded to 1 d.p.

(b) From this data, the student concluded that women have a higher resting body temperature than men. Explain why this conclusion might not be valid.
(1 mark)

The sample size is too small; there are not enough details given on how the temperatures were recorded.

(c) According to the data, does the iced water cause an increase or decrease in body temperature? Use data to support your answer. **(2 marks)**

An increase; after 10 minutes the temperature of all the subjects has increased.

(d) Some research has suggested that drinking iced water could lead to weight loss. Do these results support this idea? Explain your answer. **(2 marks)**

Yes, an increase in temperature results from an increase in metabolism, which means that more energy is being released from energy-storing molecules such as the triglycerides and glycogen.

Now try this

How can ectotherms survive for long periods without food? **(4 marks)**

If conducting an experiment to monitor physiological functions in ectotherms or endotherms you will need to consider the welfare of the organism and any ethical issues surrounding its treatment. Exposing organisms to a range of temperatures could cause them physical harm.

Excretion

Excretion refers to the removal of metabolic waste, like urea and carbon dioxide, from the body.

Metabolic waste

Any substance produced in excess by the cell, and not used, is described as metabolic waste. Getting rid of this waste is a key part of homeostasis.

- Almost every cell in the body produces carbon dioxide as a by-product of respiration.
- The liver produces nitrogenous waste (such as urea) through deamination as it breaks down excess amino acids.
- Salts, including chlorides, phosphates and sulfates, are also classified as metabolic waste products.

Carbon dioxide metabolism

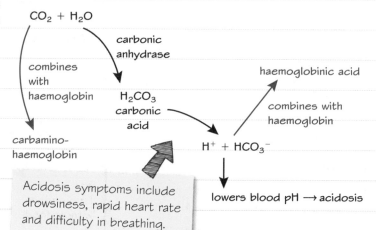

From respiration

$CO_2 + H_2O$

carbonic anhydrase

combines with haemoglobin

H_2CO_3 carbonic acid

carbamino-haemoglobin

haemoglobinic acid

combines with haemoglobin

$H^+ + HCO_3^-$

lowers blood pH → acidosis

Acidosis symptoms include drowsiness, rapid heart rate and difficulty in breathing.

Nitrogen metabolism

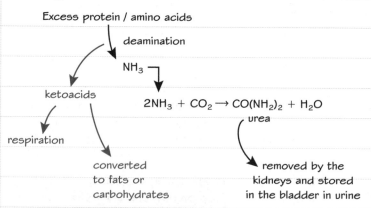

Excess protein / amino acids

deamination

NH_3

ketoacids

respiration

$2NH_3 + CO_2 \rightarrow CO(NH_2)_2 + H_2O$
urea

converted to fats or carbohydrates

removed by the kidneys and stored in the bladder in urine

The body cannot store proteins or amino acids. Potentially toxic amine groups are converted to urea, a less harmful molecule, by the ornithine cycle in the liver.

The ornithine cycle

Ammonia NH_3 CO_2 H_2O

Citrulline NH_3

Ornithine   Ornithine cycle

Arginine H_2O

H_2O

Urea $CO(NH_2)_2$

The ornithine cycle combines two molecules of ammonia and one molecule of carbon dioxide to generate one molecule of urea. Two condensation reactions and one hydrolysis reaction mean that one net molecule of water is produced in the process.

Worked example

Blockage of the airway can lead to acute acidosis. Explain how this might happen. **(4 marks)**

Carbon dioxide produced in respiration is normally removed from the blood in the lungs. If the airway is blocked then the air in the lungs cannot be replaced, a concentration gradient will not be maintained and **less** carbon dioxide will be removed from the body.

More carbon dioxide remains in the blood, forming carbonic acid. When the pH drops below 7.35, respiratory acidosis is said to have occurred.

It is important to make sure you answer the whole question and that the ideas are successfully linked. Also note the use of comparative terms (underlined). Biological processes often vary in rate rather than stopping altogether – remember to consider this in your answer too.

Now try this

List the organs involved in excretion.   **(3 marks)**

The liver – structure and function

Functions of the liver

- synthesises: bile, plasma proteins, cholesterol and red blood cells in the fetus
- stores: glycogen and vitamins
- detoxifies: alcohol and drugs
- breaks down: hormones and red blood cells
- controls blood levels of: glucose, amino acids and lipids.

Detoxification

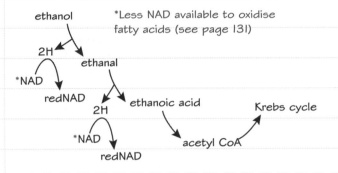

*Less NAD available to oxidise fatty acids (see page 131)

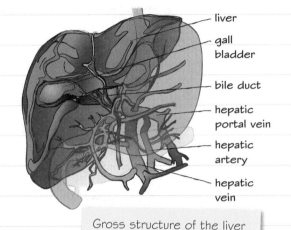

Gross structure of the liver

The liver uses the coenzyme NAD to breakdown alcohol, which plays a key role in the process of respiration (see also page 116).

Liver structure

- Branches of the hepatic artery and portal vein deliver blood to the lobule. Oxygen-rich blood from the artery and nutrient-rich blood from the portal vein pass into sinusoids.

- The liver cells process the blood before it leaves the lobule through the central hepatic vein.

- Bile collects in the bile canaliculi and drains away to the gall bladder via bile ducts, where it is stored.

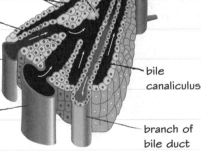

The liver is made up of functional units called lobules which are hexagonal in shape.

You need to be able to examine and draw stained sections to show the histology of liver tissue. Using a sharp pencil, draw with continuous (not sketched) lines and use the highest setting on the microscope – this is usually ×400 on a school microscope. Measurements can be made with a graticule.

A student observed the following cross-section of a mouse liver through a microscope at ×400 magnification. Label the features A–C of the liver lobule. **(3 marks)**

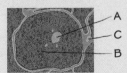

A – hepatic vein (intra-lobular vessel),
B – lobule or sinusoids,
C – inter-lobular vessel/portal vein/hepatic artery.

Now try this

If the gall bladder was removed what might be the consequences? **(2 marks)**

The kidney – structure and function

Kidneys filter the blood, removing waste such as urea and balancing the water content.

Urine production

Ultrafiltration creates the glomerular filtrate. As substances such as glucose, amino acids, salts and water are selectively reabsorbed, the concentration of urea rises. Only a small proportion of the original glomerular filtrate remains to become urine.

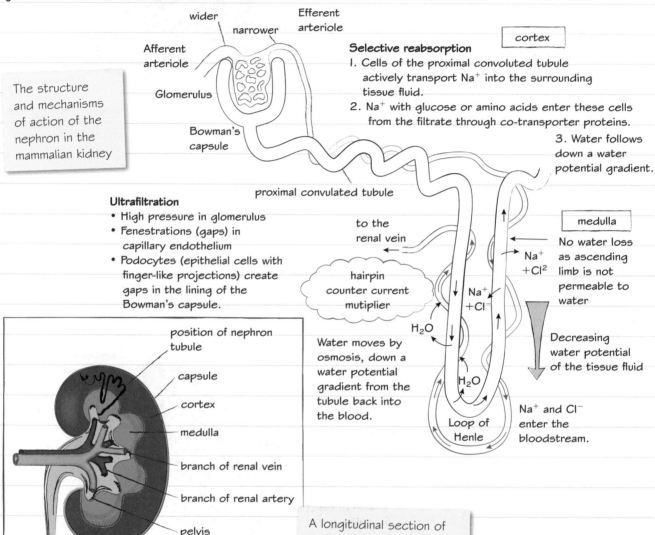

The structure and mechanisms of action of the nephron in the mammalian kidney

Selective reabsorption

1. Cells of the proximal convoluted tubule actively transport Na^+ into the surrounding tissue fluid.
2. Na^+ with glucose or amino acids enter these cells from the filtrate through co-transporter proteins.
3. Water follows down a water potential gradient.

Ultrafiltration

- High pressure in glomerulus
- Fenestrations (gaps) in capillary endothelium
- Podocytes (epithelial cells with finger-like projections) create gaps in the lining of the Bowman's capsule.

No water loss as ascending limb is not permeable to water

Decreasing water potential of the tissue fluid

Na^+ and Cl^- enter the bloodstream.

Water moves by osmosis, down a water potential gradient from the tubule back into the blood.

hairpin counter current mutiplier

A longitudinal section of the kidney showing the position of the nephron

- position of nephron tubule
- capsule
- cortex
- medulla
- branch of renal vein
- branch of renal artery
- pelvis
- ureter

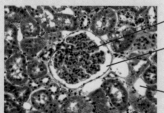

Osmoregulation

Osmoregulation is the control and regulation of the water potential of the blood and bodily fluids.

Water gains and losses

Our demands for water vary, based upon activity and external conditions.

Water gains:
- food
- drink
- metabolism (respiration).

Water losses:
- exhaled air
- urine and faeces
- sweat.

The hypothalamus and posterior pituitary

Osmoreceptor cells lose water and shrink when the water potential of the blood decreases. This stimulates neurosecretory cells to release antidiuretic hormone (ADH) from the posterior of the pituitary gland.

The pituitary gland is found on the underside of the brain just beneath the hypothalamus. The pituitary gland secretes the hormone, ADH into the blood (see page 113).

The ADH mechanism

Osmoregulation as negative feedback

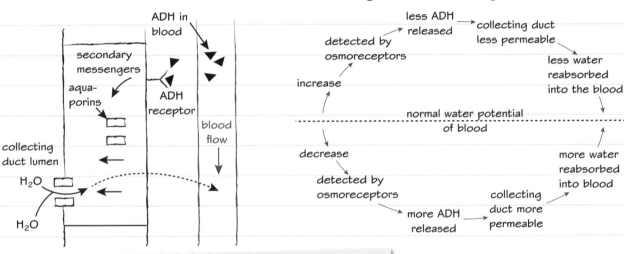

ADH arrives in the blood and binds to receptors on the surface of the cells that line the collecting duct. Aquaporins in the cytoplasm of these cells are moved into the membrane on the lumen side of the cell, making the cell more permeable to water and increasing the volume of water taken up from the filtrate.

Worked example

Caffeine is a diuretic and increases urine production. Suggest how consuming caffeine might lead to a headache. **(5 marks)**

- Caffeine increases urine production so more water is lost.
- Osmoregulatory cells detect lower blood water content.
- Neurosecretory cells do not release as much ADH.
- Less water is taken back into the blood than needed.
- Dehydration and then headaches occur.

Using bullet points to answer longer answer questions can help to keep track of the points you are making and help you earn the marks on offer.

Now try this

1　Why is it important that ADH breaks down?
　　(1 mark)

2　Suggest why aquaporins remain in the cytoplasm of the cells lining the collecting duct when not in use. **(3 marks)**

Alcohol inhibits the production of ADH and can also lead to dehydration when consumed in excess.

Kidney failure and urine testing

Causes of kidney failure

- diabetes
- hypertension
- infection.

Kidney failure results in the body being unable to remove excess water, salts and urea from the blood. The resultant imbalance of electrolytes and high concentrations of nitrogenous waste ultimately prove lethal.

Treatments for kidney failure

Dialysis involves the blood exchanging substances with a dialysis solution from which the blood is separated by a partially permeable barrier. This mimics the events in the nephron, but on a larger scale.

Transplants have advantages:

- no need for regular dialysis, one-off operation,

and disadvantages:

- lack of donors, problems with tissue matches, need to take immunosuppressant drugs.

How dialysis works

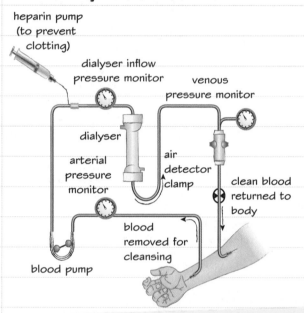

The dialyser is filled with dialysis solution which has the desired concentration of glucose and salts and no urea. Exchange occurs over 3–4 hours.

Urine testing – pregnancy

Pregnancy tests utilise two monoclonal antibodies. The first is mobile and specific to human chorionic gonadotrophin (hCG). The second is fixed at a particular point on the test strip and is specific to the hCG/antibody complex.

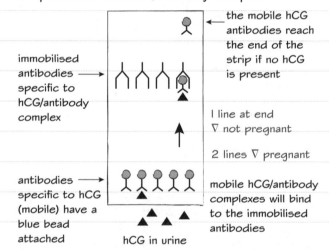

Worked example

Renal dialysis is one way of treating kidney failure. List the advantages and disadvantages of dialysis as a treatment of kidney failure. **(4 marks)**

Advantages:

- replaces the function of the kidney
- allows near-normal activity.

Disadvantages:

- time-consuming
- does not maintain the blood composition constantly.

Urine testing for drug use

Urine testing allows for the identification of a range of drugs which may have been taken by individuals. This includes testing for misuse of anabolic steroids using gas chromatography and blood alcohol levels in drivers.

Now try this

1 Suggest why dialysis solution needs to be sterile and at 37°C when it is used. **(2 marks)**

2 Sometimes a living donor can be found for a kidney transplant. Explain this. **(2 marks)**

Exam skills

This exam-style question uses knowledge and skills you have already revised. Have a look at pages 96–98 on the kidney.

Worked example

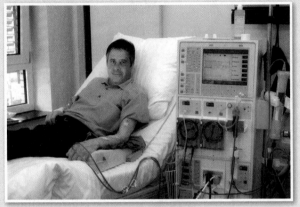

Look back at the diagram on page 98 and remind yourself of the process of dialysis.

(a) Describe what is going on in the above picture and explain why.   **(3 marks)**

The patient is having his blood filtered, to remove harmful substances including urea and to maintain the blood's water balance, by the machine on the right. This process is called dialysis. The patient must be suffering from kidney failure as the kidneys would normally carry out these processes.

(b) Under normal circumstances glomerular filtrate is formed at around 125 ml/min. Explain why we only produce 1.5 litres of urine a day.   **(3 marks)**

Most of the water is reabsorbed in the PCT, loop of Henle and the collecting duct. Sodium, glucose and amino acids that are dissolved in the glomerular filtrate are also all reabsorbed in the PCT through selective reabsorption. Only a fraction of the glomerular filtrate ends up as urine.

(c) Calculate what percentage of the glomerular filtrate ends up as urine over a 24-hour period. Give your answer to 1 dp.   **(3 marks)**

$24 \times 60 = 1440$

$1440 \times 125 = 180000$

$\dfrac{180000}{1000} = 180$ litres of filtrate in 24 hours

$\dfrac{1.5}{180} \times 100 = 0.8\%$

Command words: Describe and Explain

If a question asks you to describe something then you need to talk about what you see using scientific terms and concepts.

To explain you must talk about the underlying reasons for what you see, again using good scientific language.

Make sure you re-read exam questions after you have answered them to check that you have responded to the command words accordingly.

 Have a look back at page 96 to remind yourself of how glomerular filtrate is formed and what it consists of.

 Note the use of the correct terms for different parts of the nephron and the reabsorption process. It is important to be precise in answers to questions like this. An answer which simply said that most of the glomerular filtrate was reabsorbed would be unlikely to gain full marks.

 Maths skills Not all the information you need is immediately apparent. First you have to calculate the number of minutes in a day.

 Take care when calculating an answer that has multiple steps in it. Make a note of the numbers as you go and double-check each stage.

Always check to see if the number of decimal places required in the answer has been specified.

Sensory receptors

Sensory receptors act as transducers; detecting and converting a range of stimuli into nerve impulses.

Receptors

Type of stimuli	Type of receptor and location
Light intensity + wavelength	Photoreceptors (eye)
Sound	Mechanoreceptors (cochlea – ear)
Pressure	Mechanoreceptors (Pacinian corpuscle – skin)
Temperature	Thermoreceptors (skin)
Chemicals (volatile)	Chemoreceptors (olfactory cells – nasal cavity)
Chemicals (soluble)	Chemoreceptors (taste buds)

They all convert specific types of stimuli to generator potentials by causing gated sodium ion channels to open, creating a potential difference across the cell membrane.

Generator potentials can lead to an action potential being initiated – a nerve impulse (for more about impulse transmission, go to page 102).

Changing the potential difference across a cell membrane

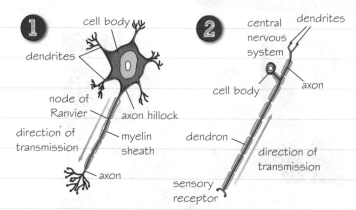

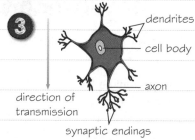

The Pacinian corpuscle

These are mechanoreceptors which respond to pressure

1. Pressure applied to layers.

2. Gated sodium ion channels open.

3. A generator potential is created.

4. If enough channels open, an action potential is produced.

How a rod cell works

opsin

ratinal

light   conformational change

change in membrane ion permeability

action potential produced

action potential

rod cell

Rod (pictured here) and cone cells, found in the retina, both contain stacks of flattened membrane sacks.

Worked example

When might a stimulus not lead to an action potential being produced?   **(4 marks)**

If the stimulus only leads to a small number of gated sodium ion channels opening, there might be insufficient change in potential difference across the membrane as the Na^+ diffuse in. Voltage-gated sodium ion channels would not open. No action potential would be produced.

Now try this

Explain why a sensory receptor can be considered to be a transducer.   **(2 marks)**

It is important to use the full term – voltage-gated sodium ion channels – as these types of channels only open with a change in the potential difference across the cell membrane.

Types of neurone

Sensory, relay and motor neurones are specialised cells that transmit action potentials and interact through synapses.

Functions of neurones

Type of neurone	Direction of nerve impulse transmission
Sensory	From receptor to central nervous system (CNS)
Relay	Between other neurones, e.g. between sensory and motor neurones, within the CNS
Motor	From the CNS to an effector, e.g. muscle or gland

Schwann cells wrap themselves around an axon or dendron many times to produce multiple layers called myelin.

The myelin sheath

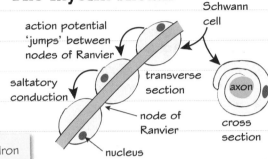

action potential 'jumps' between nodes of Ranvier

Schwann cell

saltatory conduction

transverse section

node of Ranvier

nucleus

axon

cross section

myelination speeds up a nerve impulse, e.g. from $4\,\mathrm{m\,s^{-1}}$ to $100\,\mathrm{m\,s^{-1}}$

The structures of the three types of neurone

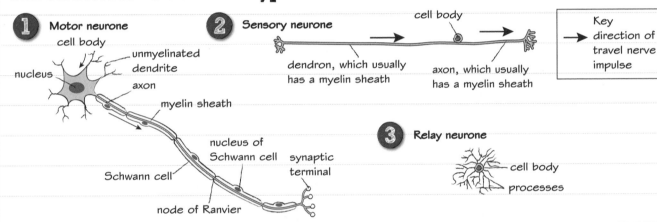

1 Motor neurone
cell body
nucleus
unmyelinated dendrite
axon
myelin sheath
Schwann cell
node of Ranvier
nucleus of Schwann cell
synaptic terminal

2 Sensory neurone
cell body
dendron, which usually has a myelin sheath
axon, which usually has a myelin sheath

Key
→ direction of travel nerve impulse

3 Relay neurone
cell body
processes

Neurones (or nerves cells) are regular cells that have nuclei, rough ER, Golgi bodies and cytoplasm. They differ in the location of the cell body, axon length/presence and myelination. Around one third of peripheral neurones are myelinated. The remainder (and the neurones in the CNS) are unmyelinated. Myelinated neurones tend to be longer, carrying signals quickly over long distances.

Worked example

There are similarities and differences between the structures of motor and sensory neurones. Complete the table below.

(4 marks)

	Type of neurone	
Feature	**Sensory neurone**	**Motor neurone**
Position of the cell body	Just outside the CNS	Inside the CNS
Is it myelinated?	Yes	Yes
Direction of communication	From receptor to CNS	From the CNS to an effector e.g. muscle or gland
Length of axon	Short – runs from just outside the CNS into the CNS	Long – runs from inside the CNS to the effector

Now try this

1 How does myelination affect transmission of a nervous impulse? **(1 mark)**
2 Explain the difference between the axon and the dendron of a sensory neurone. **(2 marks)**
3 Are neurones likely to contain mitochondria? Explain why. **(2 marks)**

It is easy to confuse the axon and the dendron of a sensory neurone – they are named in relation to the location of the cell body, not in terms of what they look like.

Action potentials and impulse transmission

An action potential is a depolarisation of the cell membrane so the inside becomes more positive than the outside; the transmission of an action potential along a neurone is a nerve impulse.

The resting potential

This is maintained at about $-60\,mV$ inside the cell compared to outside the cell.

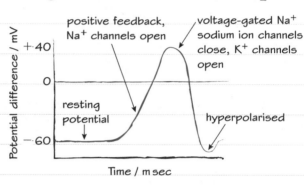

Na$^+$/K$^+$ pumps use ATP to pump 3 Na$^+$ out of the cell whilst taking 2 K$^+$ in. As some K$^+$ channels are open at rest, some flow out down a concentration gradient – hence, a relatively negatively charged environment is created inside the cell.

The generation of an action potential

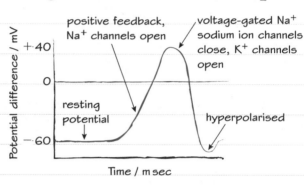

Intensity of a stimulus

When a stimulus has a high intensity a large number of action potentials are generated and it is these that our brain interprets as, for instance, a bright light or loud noise.

Nerve impulses only travel in one direction

When a region of the neurone is hyperpolarised, no further action potentials can be produced – a refractory period. This is because the Na$^+$ and K$^+$ need to be re-distributed so that a resting potential can be re-established.

Local currents

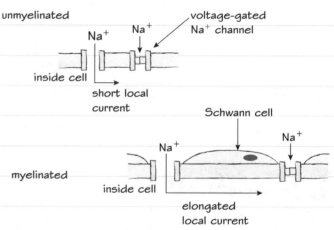

Worked example

Nerve impulses travel more quickly along myelinated neurones. Explain how the myelin sheath speeds up a nerve impulse. **(4 marks)**

The myelin sheath **insulates** the neurone. Na$^+$ and K$^+$ can only pass across the membrane in the gaps between the **Schwann cells** – **nodes of Ranvier**. This leads to the action potentials jumping from one node to the next – **saltatory conduction**.

These key terms need to be used in your answer. Making a list of key terms next to the question can help when it comes to giving your answer.

Now try this

1 What causes a voltage-gated sodium ion channel to open?
(2 marks)

2 Why is it impossible to stimulate another action potential when a cell membrane is hyperpolarised?
(3 marks)

Structure and roles of synapses

Synapses are junctions between neurones where chemicals called neurotransmitters are used to pass on action potentials.

The structure of the cholinergic synapse

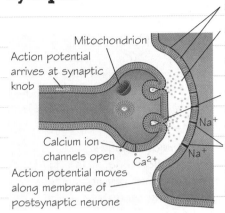

Sodium ion channels

Acetylcholine molecules diffuse across cleft and attach to receptors on sodium ion channels

Mitochondrion

Action potential arrives at synaptic knob

Vesicles of acetylcholine fuse to presynaptic membrane

Na⁺

Sodium ion channel opens to allow sodium ions into the postsynaptic neurone

Calcium ion channels open

Ca²⁺

Na⁺

Action potential moves along membrane of postsynaptic neurone

The action of neurotransmitters

1 Action potential arrives at synaptic knob.

2 Calcium-gated channels open and calcium ions diffuse in.

3 Calcium ions cause neurotransmitter vesicles to fuse with the pre-synaptic membrane.

4 Acetylcholine diffuses across the synaptic cleft to bind with receptors on sodium ion channels.

5 Sodium ion channels open and a generator potential is created (EPSP – excitatory post-synaptic potential).

6 If the EPSP is of sufficient magnitude an action potential is produced.

The importance of synapses

Synapses have a number of important roles in summation and control.

- ✓ Synapses allow for the convergence of signals from many neurones to one neurone

- ✓ Synapses allow for the divergence of signals from one neurone to many neurones.

They also:

- ✓ ensure impulses travel in one direction only

- ✓ filter out low-level stimuli

- ✓ amplify low-level stimuli

- ✓ allow new (neuronal) pathways to occur – the basis for conscious thought and memory.

Summation

Summation refers to the way several small potential changes can combine to produce one large change in potential difference. It can be temporal or spatial and allows low-level stimuli to be amplified.

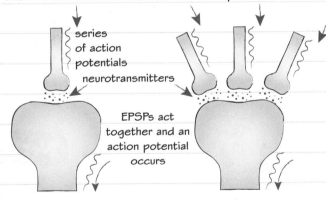

series of action potentials neurotransmitters

EPSPs act together and an action potential occurs

Temperal summation

Spatial summation

Now try this

1 Mitochondria are found in large numbers in synaptic knobs, suggest why.   **(2 marks)**
2 Why is it important that the synaptic cleft is a very narrow gap?   **(2 marks)**
3 Which three types of ion play particularly important roles in nerve impulse transmission?   **(3 marks)**

Endocrine communications

Endocrine glands release hormones into the blood where they act as messengers, activating specific target organs or tissues.

Glands

There are two types of gland in the body:

- **Endocrine glands** secrete hormones directly into the blood. The blood transports the hormones around the body so they can reach target cells.

- **Exocrine glands** do not release hormones. They have small tubes (ducts) which carry their secretions to other places. For example, the secretion of saliva in the mouth.

	Endocrine	Exocrine
Secretion	Hormones	Other molecules including enzymes and bile
Duct	No	Yes
Destination	The blood	Other parts of the body, e.g. digestive system

Target cells

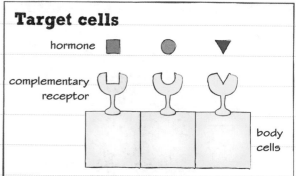

If a cell has a receptor which is complementary to a hormone, it is a target cell and can respond appropriately. Cells will normally have many receptors which allow them to detect different hormones.

Endocrine glands

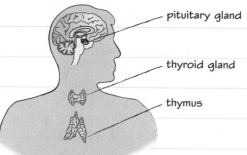
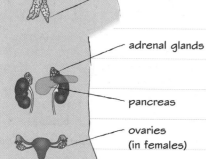

- pituitary gland
- thyroid gland
- thymus
- adrenal glands
- pancreas
- ovaries (in females)
- testes (in males)

The pituitary gland produces a number of hormones including antidiuretic hormone (ADH), follicle-stimulating hormone, luteinising hormone and oxytocin.

How hormones work

Hormones bring about a response inside the cell when they bind to specific membrane-bound receptors. Sometimes this involves a second messenger such as cyclic AMP, which activates enzymes to alter the behaviour or output of the cell.

Steroid hormones are lipid-soluble hormones which can pass through the cell membrane, interact with other proteins in the cytoplasm and affect the transcription of target genes directly.

> Used hormones break down naturally or are filtered out in the kidneys and lost in urine.

Worked example

Endocrine glands secrete hormones that are often some distance from their target tissues. Explain why target cells must have 'specific' receptors. Use an example in your answer. **(3 marks)**

ADH is produced in the pituitary gland but it acts on the cells of the collecting duct within the nephrons of the kidney. Only those cells have complementary receptors; otherwise, ADH would bind to other cells around the body and not be available to bring about an increase in the permeability of the collecting ducts.

> Using examples will help you to give specific details.

Now try this

1 How do hormones reach target organs? **(1 mark)**
2 Explain the role of second messenger molecules. **(2 marks)**

Endocrine tissues

Adrenal glands are an example of endocrine glands; they produce and secrete adrenaline, which acts on a wide range of target tissues, preparing the body for action. The pancreas also contains endocrine tissue.

Role of adrenaline

Adrenaline binds to a large number of different but complementary receptor types bringing about a range of different processes. Examples include:

- ✓ conversion of glycogen to glucose
- ✓ dilation of pupils
- ✓ increase in heart rate and stroke volume
- ✓ relaxation of smooth muscle in bronchioles
- ✓ inhibition of gut action
- ✓ the erection of body hair.

The structure and function of the adrenal gland

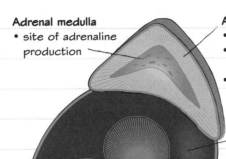

Adrenal medulla
- site of adrenaline production

Adrenal cortex
- produces certain steroid hormones
- mineral corticoids – balance Na^+ and K^+ in blood
- glucocorticoids – control carbohydrate and protein metabolism in the liver

kidney cortex

kidney medulla

Adrenaline is released when the adrenal gland is innervated by a motor neurone from the CNS.

Histology of the pancreas

The pancreas contains both endocrine and exocrine tissue. The islets of Langerhans contain α and β cells which secrete glucagon and insulin respectively – this is the endocrine part of the pancreas.

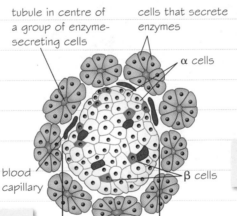

tubule in centre of a group of enzyme-secreting cells

cells that secrete enzymes

α cells

blood capillary

β cells

The α and β cells in islet of Langerhans

islet of Langerhans

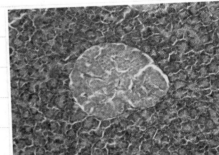

A section through the pancreas showing an islet of Langerhans (×100)

Worked example

Smoking cigarettes leads to an increase in adrenaline levels. Explain how smoking for many years might affect someone's health due to the effect of adrenaline.   **(3 marks)**

A person who smoked may have a higher resting heart rate and stroke volume due to the increase in adrenaline. This could lead to a long-term increase in blood pressure, which could cause damage to blood vessels and lead to complications such as coronary heart disease or strokes.

Now try this

The adrenaline–receptor complex changes conformation when adrenaline binds. What does this mean?

(2 marks)

Regulation of blood glucose

Blood glucose concentration is regulated by the actions of insulin and glucagon. Insulin is released in response to elevated glucose levels in the blood.

Regulating the concentration of glucose in the blood

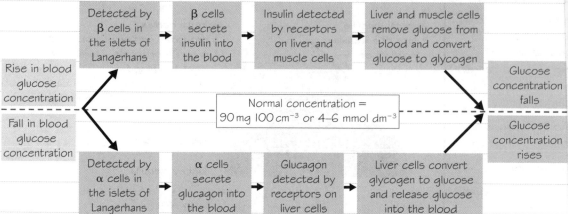

This diagram shows the role of the liver, insulin and glucagon in regulating blood glucose. This is an example of negative feedback and a homeostatic mechanism.

Rise in blood glucose concentration → Detected by β cells in the islets of Langerhans → β cells secrete insulin into the blood → Insulin detected by receptors on liver and muscle cells → Liver and muscle cells remove glucose from blood and convert glucose to glycogen → Glucose concentration falls

Normal concentration = 90 mg 100 cm^{-3} or 4–6 mmol dm^{-3}

Fall in blood glucose concentration → Detected by α cells in the islets of Langerhans → α cells secrete glucagon into the blood → Glucagon detected by receptors on liver cells → Liver cells convert glycogen to glucose and release glucose into the blood → Glucose concentration rises

Mechanism of insulin secretion

1 The cell membrane has potassium and calcium ion channels

2 The potassium ion channels are normally open – so potassium ions flow out

3 When blood glucose concentration is high the glucose moves into the cell

8 Calcium ions cause the vesicles of insulin to fuse with cell membrane, releasing insulin by exocytosis

Insulin is secreted by β cells in the islets of Langerhans.

Glucose
Glucokinase
Glucose phosphate
ATP

4 Glucose is metabolised to produce ATP

5 The ATP closes the potassium ion channels

7 The change in potential difference opens the calcium ion channels

6 The accumulation of potassium ions alters the potential difference across the cell membrane – the inside becomes less negative

Worked example

As you wake up after a long night's sleep, will glucagon or insulin be most likely to be required in higher concentrations? Explain why and explain what effect the hormone will have.   **(3 marks)**

Glucagon. The levels of glucose in the blood are likely to be low as it has been a long time since the last meal. Glycogen reserves in the liver can be converted to glucose and released into the bloodstream and the blood glucose concentration will increase.

Now try this

1 Where will you find α and β cells?   **(2 marks)**
2 Insulin secretion has events in common with a nerve synapse. Outline the role of Ca^{2+} in each.
    (3 marks)

Most pancreatic cells secrete digestive enzymes (exocrine function).

Islets of Langerhans in the pancreas contain α and β cells that detect decreases and increases in glucose concentration and release glucagon or insulin respectively.

Diabetes mellitus

Diabetes mellitus is a disease in which the body is unable to control blood glucose concentration effectively.

Type I diabetes

- starts in childhood
- sufferer cannot manufacture sufficient insulin
- cause: strong genetic component, linked to autoimmune attack on β cells

Type II diabetes

- starts later in life, normally 50+
- pancreas still produces insulin but there is a decline in response to insulin so glucose levels remain high for longer periods
- causes: linked to obesity and high-sugar diet; more likely in people of Asian or Afro-Caribbean origin.

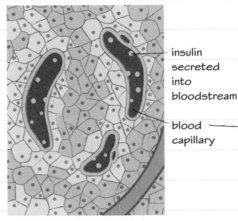
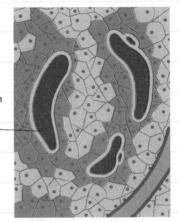

insulin secreted into bloodstream

blood capillary

☐ insulin-producing cells

■ insulin-producing cells destroyed

T-cells incorrectly target β cells for destruction

Treatments

monitor blood glucose concentration

regular insulin injections

Type I

Treatment

Type II

Hypoglycaemia – blood glucose level too low.

Hyperglycaemia – blood glucose level too high.

monitor and control diet

lower carbohydrate intake

may involve insulin injections at later stages

Sources of insulin

Insulin used to be obtained from animals such as pigs. **Bacteria** engineered to contain the human insulin gene can now be used as an abundant source of human insulin. This modern approach is better because:

- ✓ lower risk of infection
- ✓ cheaper and easier to manufacture
- ✓ fewer moral objections.

Stem cells potentially offer the possibility of replacing lost or damaged β cells with previously undifferentiated cells, which could become specialised into new functional β cells when placed in the right environment.

Further consequences of diabetes include:

- eye damage
- nerve damage
- kidney damage
- strokes and heart attacks.

Worked example

Explain why a symptom of type II diabetes might be high blood pressure. **(3 marks)**

Elevated glucose concentrations in the blood mean the water potential of the blood would drop. This could lead to an increase in water uptake from the collecting ducts of the kidney nephrons and a subsequent increase in blood pressure.

Now try this

1 Why might fewer people have moral objections about the modern-day production of insulin? **(2 marks)**

2 There are a number of 'risk factors' associated with type II diabetes. Explain what is meant by this statement.

(2 marks)

Plant responses to the environment

Plants use a range of growth responses to deal with abiotic stress (stress from a non-living source) and herbivory (the eating of plants).

Tropism

A **tropism** is a directional growth response to stimuli. It can be defined as being positive (towards) or negative (away).

- Phototropism – towards (+ve) or away (−ve) from light.
- Geotropism – towards (+ve, i.e. downwards) or away (−ve, i.e. upwards) from the pull of gravity.
- Chemotropism – a pollen tube grows towards ovules in the ovary (+ve).

Mimosa pudici responds to touch by drooping to avoid wind damage. Non-directional responses like this are called **nastic** responses.

Stomatal closure – a response to stress

Normal

open stoma, gas exchange occurs

Stress response

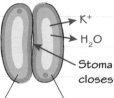

K⁺
H₂O

Stoma closes

a lack of water in the roots leads to ABA entering guard cells

water loss leads to guard cells losing turgidity stoma closes

Photosynthesis relies on open stomata for gas exchange to occur. Rates of photosynthesis are therefore reduced during times of stress caused by a lack of water.

Further roles of plant hormones

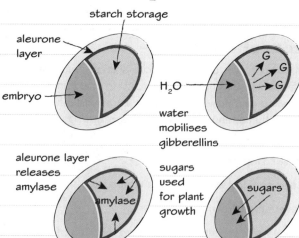

starch storage

aleurone layer

embryo

H_2O →

water mobilises gibberellins

G G G

aleurone layer releases amylase

amylase

sugars used for plant growth

sugars

Seed germination: Abscisic acid (a hormone) represses the transcription of the amylase gene, but high levels of the hormone gibberellin activate transcription to amylase and can begin to break down the seed's starch reserves.

photosynthesis products leaving the leaf

xylem

phloem

abscission layer

protective layer – waterproofs at abscission

Leaf abscission (loss)

1 A leaf's pigment begins to break down in autumn, its auxin levels also fall.

2 In the absence of auxin the abscission zone becomes sensitive to another hormone – ethene.

3 Ethene increases the production of cellulose, which acts on the cellulose of the cell walls in the abscission zone.

4 The leaf falls from the plant.

Worked example

Some plants release alarm pheromones when they are grazed on, which cause tannins and other chemicals such as alkaloids to be made in neighbouring plants. Suggest why this might occur. **(3 marks)**

It would make the plant very bitter and drying to eat, so herbivores would not eat those particular plants. This would help other members of the same species to survive in that location.

Now try this

1 What role does water play in seed germination? **(1 mark)**

2 Suggest why a mutation in the gene that codes for cellulase could prevent leaf abscission. **(2 marks)**

Controlling plant growth

Experimental evidence, arising from many experiments, has allowed us to understand the hormonally controlled mechanisms behind apical dominance and stem elongation.

Auxins and apical dominance

Observation 1: Removal of the apex of a plant leads to lateral bud growth.

Observation 2: Auxin paste applied to the cut end and no lateral bud growth.

Observation 3: Auxin transport inhibitors stop auxin moving down the stem and lateral bud growth occurs.

Conclusion: High levels of auxin stop lateral bud growth.

— Based on the experimental evidence at this point, the conclusion is correct.

Observation 4: Abscisic acid inhibits bud growth.

Observation 5: Cytokinins promote bud growth.

Observation 6: High concentrations of auxin sequester cytokinins.

Observation 7: Auxin levels in lateral buds increase when the apex is cut off.

This observation appears to conflict with our original conclusion, so a new conclusion must be reached based on the evidence given.

New conclusion: High concentrations of auxin in the apex maintain high concentrations of abscisic acid. When the apex is removed auxin levels drop, cytokinins are released and spread out, and lateral bud growth occurs.

All observations can be explained in the context of this conclusion; therefore, this theory stands until new evidence disproves it. There is no direct causative link between auxins and apical dominance.

Gibberellins and stem elongation

Observation 1: Dwarf pea plants grow taller when gibberellic acid (GA3) is applied to the stem.

Conclusion: GA3 is responsible for plant stem growth.

— This does not prove a causal link: other substances may be involved; the GA3 concentration may have been artificially high.

Observation 2: Tall pea plants express the LE enzyme and contain high levels of another gibberellin – GAI.

Observation 3: Dwarf pea plants do not express the LE enzyme and have lower levels of GAI.

Conclusion: The LE enzyme and GAI have roles in stem elongation.

— How are the two substances involved in stem elongation?

Observation 4: A pea plant with a mutation in the GA pathway (see diagram below) but functional LE enzyme produces no GA and grows to just 1 cm.

Observation 5: Grafting a shoot from this plant onto a dwarf pea plant that did not produce LE enzyme (homozygous le) led to normal growth.

New conclusion: The LE enzyme is required to convert the precursor of GAI into GAI. LE diffused from the grafted shoot into the dwarf plant and then catalysed the production of GAI. This confirms that GAI causes stem elongation.

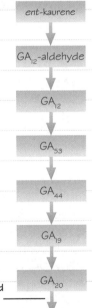

ent-kaurene

GA$_{12}$-aldehyde

GA$_{12}$

GA$_{53}$

GA$_{44}$

GA$_{19}$

GA$_{20}$

Enzyme produced by the **LE** allele acts here —

GA$_1$

Now try this

1 Explain why it would be incorrect to say that a lack of auxin causes lateral bud growth in plants. **(2 marks)**

2 Gibberellins stimulate internodal growth. For growth in plants to occur, suggest two things that gibberellins might affect. **(2 marks)**

Plant responses

🧪 **Practical skills** Experiments have shown how plants respond to light. Serial dilutions can be used to show how plant hormones affect growth.

Phototropism

These experiments have helped us to understand the mechanism behind phototropism.

1 The absence of tip shows that the response to light depends on its presence.

2 The covered tip shows the need for light to reach the tip for a response to occur.

3 The transparent cover reinforces point 2.

4 The opaque shield at base has no effect, confirming the role of the tip.

5 The gelatine block allows chemicals to flow from the tip into the rest of the plant.

6 The mica prevents the flow of chemicals.

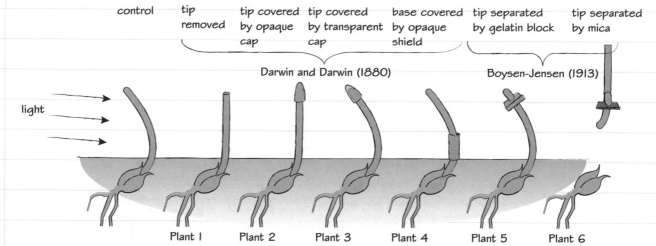

control | tip removed | tip covered by opaque cap | tip covered by transparent cap | base covered by opaque shield | tip separated by gelatin block | tip separated by mica

Darwin and Darwin (1880)   Boysen-Jensen (1913)

light

Plant 1   Plant 2   Plant 3   Plant 4   Plant 5   Plant 6

Worked example

A range of dilutions of commercially available auxin were used to produce agar blocks. The concentration of the stock solution was 1 M. Complete the table below to show how serial dilutions were used to create a range of auxin concentrations. **(4 marks)**

	Solution used to make agar blocks				
Tube	1	2	3	4	5
Auxin solution (cm³)	1.0	0.1 of tube 1	0.1 of tube 2	0.1 of tube 3	0.1 of tube 4
Water (cm³)	0.0	0.9	0.9	0.9	0.9
Auxin concentration (M)	1.0000	0.1000	0.0100	0.0010	0.0001

What concentration of auxin most closely matches the concentration produced by the tip under normal growth? Explain your answer. **(2 marks)**

0.01 M, as the growth pattern most closely matches that of the plant that has not had the tip removed.

The investigation was repeated using just plants 5 and 6. After 14 days the following results were recorded:

	Total lateral bud growth, measured distance from main stem to tip of first lateral bud (mm)							
	Day							
Plant	0	2	4	6	8	10	12	14
1	0	0	0	0	0	0	0	0
2	0	0	0	0	0	0	1	3
3	0	0	0	0	1	2	4	6
4	0	1	2	4	5	8	10	12
5	0	1	4	6	8	13	17	22
6	0	2	5	7	9	14	19	25
7	0	0	0	0	1	2	3	4

Commercial use of plant hormones

Plant hormones are used in a variety of different ways to enhance productivity in the food and drink industry. Examples are rooting powders, controlling ripening and killing weeds.

Uses of auxins

- Dipping the end of a cutting in rooting powder containing auxins before planting encourages root growth thanks to the presence of meristem cells in the cutting.

- Weedkillers containing auxins cause broad-leafed weeds to grow more rapidly than their root systems can support. Narrow-leafed plants such as grasses have a lower surface area and are not affected in the same way.

Use of cytokinins and ethane

Cytokinins

- delay yellowing of picked lettuce
- promote bud and shoot growth in micropropagation.

Ethene

- can be inhibited by cool temperatures and low O_2 levels to avoid early ripening
- promotes fruit drop (cherry, cotton)
- speeds up fruit ripening (apples, tomatoes).

Uses of gibberellins

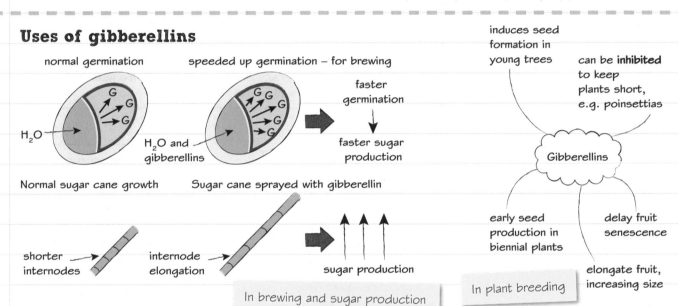

Worked example

Outline how plant hormones could be used in the production of a cherry-flavoured fruit beer. **(3 marks)**

Gibberellins: Speed up the malting process so more sugar is made more rapidly by the germinating seeds.

Ethene: Promote fruit drop in cherries.

The use of plant hormones in fruit growth and transportation is widespread and contributes significantly to the profits made in this industry.

Make sure you remember the role of auxins in rooting powders and weedkillers.

Now try this

1. Keeping unripe avocados with bananas is thought to speed up the ripening of the avocados. Suggest why. **(1 mark)**

2. How can a herbicide that speeds up plant growth actually work as a weedkiller? **(3 marks)**

Mammalian nervous system

In animals, responding to changes in the environment is a complex process. The nervous system enables effective communication between sensors and effectors, leading to a coordinated response to these changes.

Nervous system

Central nervous system: made up of grey (billions of unmyelinated neurones) and white (longer myelinated neurones) matter

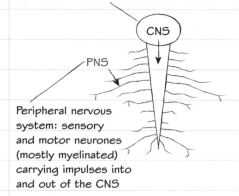

Peripheral nervous system: sensory and motor neurones (mostly myelinated) carrying impulses into and out of the CNS

The functional organisation of the nervous system

Receptors and sensory neurones transmit impulses to the CNS. Motor neurones then carry out somatic or autonomic functions.

Somatic:

✓ motor neurones transmit impulses to voluntary muscles

✓ mostly myelinated.

Autonomic:

✓ motor neurones carrying impulses to involuntary muscles and glands

✓ self-governing; controls the majority of homeostatic mechanisms, largely subconscious, e.g. cardiac muscle, smooth muscle in wall of gut.

The autonomic nervous system

This consists of parasympathetic and sympathetic motor neurones

Parasympathetic (rest and digest):

- acetylcholine is the neurotransmitter between the neurone and the effector
- ganglion is in target tissue
- reduces heart rate
- sleep and relaxation
- reduces ventilation
- pupil constriction
- sexual arousal.

Sympathetic (fight or flight):

- noradrenaline is the neurotransmitter between the neurone and the effector
- ganglion just outside spinal cord
- increases heart rate
- upregulated during stress
- increased ventilation rate
- pupil dilation
- orgasm.

The autonomic nervous system can be sub-divided into two antagonistic systems. At rest, impulses pass along neurones of both systems. Changes in the environment dictate a shift in balance; that is, one will be upregulated relative to the other.

Worked example

Whilst waiting to cross the road, the lights in front of you turn green and you begin to walk. To your right you hear the screeching of tyres as a car only just stops before the crossing.

Outline how the different parts of the nervous system enable you to cross the road and suggest how you might respond to the near incident with the car. **(6 marks)**

Initially, photoreceptors in the eye send impulses through sensory neurones to the CNS.

The somatic system sends impulses from the CNS via motor neurones to the muscles to allow you to walk across the road.

Receptors in the ear pick up the sound of the car braking; impulses pass through sensory neurones to the CNS. The sympathetic part of the autonomic nervous system is upregulated so the heart rate and ventilation rate increase and pupils dilate.

Now try this

1 The sympathetic nervous system is sometimes linked to the 'fight or flight' response. Why? **(3 marks)**

2 What is meant by the term 'autonomic' nervous system? **(1 mark)**

 Longer answer questions require planning. Note how this answer is set out chronologically, so each part can be linked to nervous system activity.

The brain

The brain is a concentrated mass of nervous tissue which acts as the coordination centre for all conscious and subconscious activity in the body.

The structure of the human brain

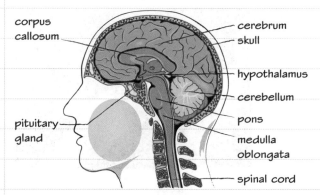

corpus callosum
cerebrum
skull
hypothalamus
cerebellum
pons
medulla oblongata
pituitary gland
spinal cord

A highly developed cerebrum is indicative of a mental sophistication found in higher mammals. In humans, this part of the brain has developed considerably over time.

The roles of the cerebrum

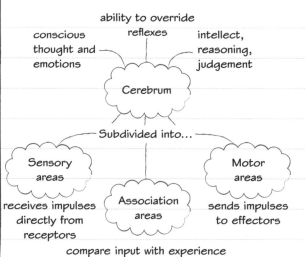

ability to override reflexes
conscious thought and emotions
intellect, reasoning, judgement

Cerebrum

Subdivided into...

Sensory areas
Association areas
Motor areas

receives impulses directly from receptors
compare input with experience to enable appropriate response
sends impulses to effectors

Other important parts of the brain

Medulla oblongata:
- control of non-skeletal muscle
- respiratory centre
- cardiac centre.

Hypothalamus:
- osmoregulatory centre
- thermoregulatory centre
- regulates pituitary gland.

Pituitary gland

A number of key hormones are released from the pituitary gland, including Antidiuretic hormone (ADH) Growth hormone (GH), Adrenocorticotropic hormone (ACTH) and Thyroid stimulating hormone (TSH).

The function of the cerebellum

The body's 'autopilot', the cerebellum, coordinates input from a variety of sense organs ensuring the fine control of muscular movements involved in balance, movement and tool use.

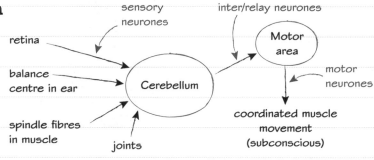

sensory neurones
inter/relay neurones
retina
Motor area
balance centre in ear
Cerebellum
motor neurones
spindle fibres in muscle
joints
coordinated muscle movement (subconscious)

Now try this

1 Which part of the brain contains the cell bodies of the motor neurones which control heart rate?
(1 mark)

2 Damage to the cerebrum might alter someone's personality but not lead to death. Explain why.
(2 marks)

The brain is highly complicated and sophisticated. Focus on learning the structures named here and their basic functions. Importantly, you need to be able to relate the brain region activities to the structure of neurones and the roles of synapses.

Reflex actions

Reflex responses are simple forms of genetically determined behaviour which are carried out subconsciously and often have a protective purpose.

A reflex arc

A reflex is a rapid, involuntary response to a stimulus. It involves an impulse passing along a sensory neurone, via an inter-neurone in the CNS directly to a motor neurone. The response is automatic and unconscious.

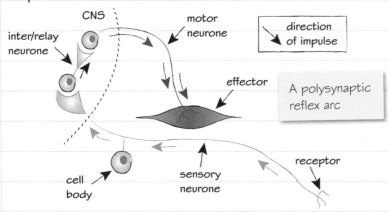

A polysynaptic reflex arc

The blinking reflex

Touch on the cornea is detected by a sensory neurone receptor. Coordination takes place in the pons of the brain stem. A motor neurone stimulates the muscles around the eye and the blink response occurs.

The knee jerk reflex

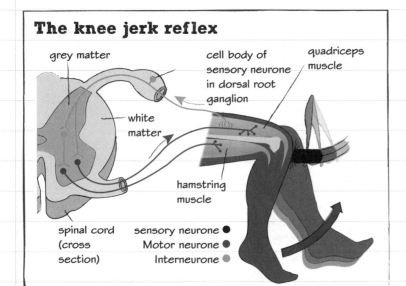

Survival value

Primitive reflexes enable survival in new-borns:

- rooting reflex – searching for the nipple
- sucking reflex
- grasping reflex
- swimming reflex.

These types of reflexes in humans help to ensure feeding and other life-preserving processes occur, aiding survival for the initial period following birth.

Worked example

A student conducted a simple experiment looking at response times. A ruler was held above the open hand of each subject and dropped. The distance the ruler dropped was recorded:

Subject	Distance ruler drops before being caught (mm)					Mean	Standard deviation
A	45	55	60	48	63	54.2	7.66
B	55	60	65	66	58	60.8	4.6

Complete the table by filling in the mean and standard deviation for each subject. **(4 marks)**

Sample working: standard deviation calculated for subject A.

$(45 - 54.2)^2$	-9.2^2	84.64
$(55 - 54.2)^2$	0.8^2	0.64
$(60 - 54.2)^2$	5.8^2	33.64
$(48 - 54.2)^2$	-6.2^2	38.44
$(63 - 54.2)^2$	8.8^2	77.44
	Total	234.80

$$s = \sqrt{\frac{\Sigma(x - \bar{x})^2}{n - 1}}$$

$$\frac{234.80}{4} = 58.7$$

$$\sqrt{58.7} = 7.66$$

Coordination of responses

The 'fight or flight' response to environmental stimuli involves a coordinated response by the nervous and endocrine systems.

Coordination of the fight or flight response

The fight or flight response in animals is particularly noticeable in predators, where the bearing of teeth, flattening of ears, growling and raising of hair accompany the other less visible responses to stress / threat.

A general discharge of the sympathetic nervous system enables fight or flight through the coordination of nervous and hormonal effectors.

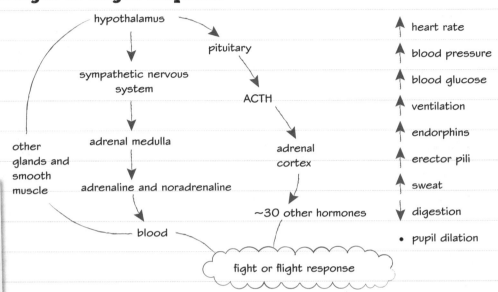

↑ heart rate
↑ blood pressure
↑ blood glucose
↑ ventilation
↑ endorphins
↑ erector pili
↕ sweat
↓ digestion
• pupil dilation

The action of adrenaline as a first messenger

The binding of adrenaline to its receptor on the cell surface membrane results in the generation of cyclic AMP. This second messenger initiates further reactions inside the cell.

The result of this process will depend on the type of cell and the type of receptor present. Adrenaline can cause vasoconstriction of some blood vessels (gut) and vasodilation of others (muscle).

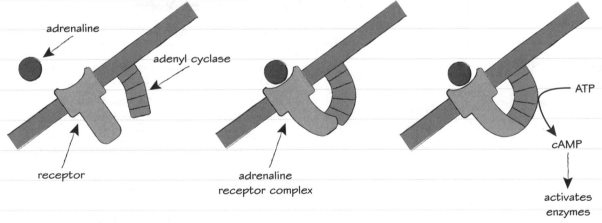

You may have heard about PTSD before. Use your own knowledge gained beyond the syllabus. For instance, people with PTSD can suffer flashbacks and periods of elevated stress leading to depression.

Worked example

Post-traumatic stress disorder (PTSD) is linked to long-term exposures to the stresses of warfare and the repeated stimulation of an inappropriate fight or flight response. Suggest what type of long-term health issues might arise from this. **(3 marks)**

Digestive problems, high blood pressure and associated cardiovascular issues and anxiety. These can all be linked to chronically high levels of adrenaline.

Now try this

Why might the feeling of stress or panic affect our digestion? **(2 marks)**

115

Controlling heart rate

Nervous and hormonal mechanisms combine to control heart rate.

Control of heart rate

At rest the sinoatrial node (SAN) maintains the heart rate at 60–80 beats per minute.

The frequency of the waves of excitation that spread out from the SAN is controlled by the cardiovascular centre in the medulla oblongata.

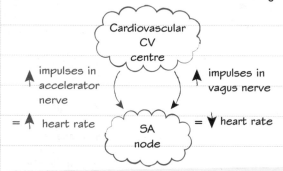

Impulses via the vagus nerve slow the heart rate, while impulses via the accelerator nerve increase the heart rate.

Affecting the heart rate

The cells that make up heart tissue have adrenergic receptors. **Adrenaline** binds to these receptors triggering a range of reactions inside the cells, the net effect of which is to speed up heart rate.

Pacemakers are devices that deliver an electrical impulse to heart muscle. They can be used in people with problems with their SAN.

🧪 Practical skills — Mean Values

Person	Pulse rate (BPM)		
	At rest	After exercise	Pulse increase (BPM)
1	70	105	35
2	67	98	32
3	65	120	55
4	58	105	47
5	78	110	32
6	80	135	55
7	56	90	34
8	60	96	36
9	65	102	37
10	70	99	29
Mean	66.9	106.0	39.1
St Dev	7.94	13.08	5.14

Look at the mean values at rest and after exercise – there is a clear difference. Exercise causes an increase in heart rate.

Responding to feedback

Receptors in various places around the body send impulses to the cardiovascular centre.

blood pressure monitored by stretch receptors in carotid arteries

stretch receptors in muscles

↓ pH

CV centre

chemoreceptors in aorta, carotid artery and brain detect changes in blood pH

↑ pH

increase in heart rate

decrease in heart rate

Stimuli that lead to an increase in heart rate have red arrows. Blue arrows lead to a decrease.

This is a short question but it carries a lot of marks. Plan your answer carefully and make sure you have read and understood exactly what you are being asked.

Worked example

Explain the response of the heart to exercise.
(6 marks)

Exercise brings about a higher rate of respiration. This leads to an increase in carbon dioxide in the blood and a subsequent decrease in pH (as carbon dioxide reacts with water in the blood to make carbonic acid). Chemoreceptors in the aorta and carotid arteries send impulses to the cardiovascular centre.

An increase in the number of action potentials sent down the accelerator nerve occurs and more waves of excitation are sent from the SAN across the heart to the atrioventricular node (AVN). The heart rate increases.

Now try this

1 The SAN is sometimes called the body's pacemaker. Why?   **(1 mark)**
2 How does carbon dioxide lead to a decrease in the pH of the blood?   **(2 marks)**

Muscle structure and function

Types of muscle

Types of muscle	Cell structure	Location	Function
Cardiac muscle cell	Myogenic, rows of cells connected through intercalated discs, striated	3 types: atrial, ventricular and conductive muscle fibres; all in the heart	Consists of atrial and ventricular muscle, which contract like skeletal muscle and excitatory and conductive muscle fibres, which conduct electrical impulses and control the heartbeat. Some cardiac muscle is myogenic – can contract without nervous input.
Skeletal muscle cell	Form fibres, multinucleate, many mitochondria, contain contractile elements called myofibrils, cell membrane called sarcolemma	All skeletal muscle	Allow for the movement of the skeleton at joints, they work in antagonistic pairs
Smooth muscle cell (involuntary)	Spindle-shaped, around 500 µm in length when relaxed and 5 µm wide	Walls of intestine, iris of eye, walls of arteries and arterioles	Found in the walls of arteries and arterioles and walls of intestine and uterus

Skeletal muscle is arranged in bundles of fibres that contain myofibrils sub-divided into functional units called sarcomeres (between two Z-lines).

The sliding filament model

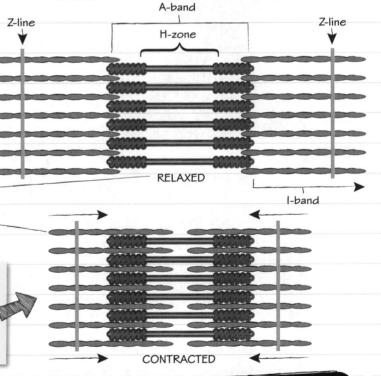

Myosin – thick protein filament, each molecule consisting of a tail and two head regions.

Actin – thin filament of globular subunits, twisted like a double strand of beads.

The role of ATP in contraction is covered on page 118.

All muscle cells produce a force on contraction because they contain filaments made of the proteins actin and myosin.

The role of ATP in contraction is covered on page 118.

Worked example

Muscles only exert force in one direction. Explain how this issue is overcome in:

(a) vasoconstriction in arterioles **(2 marks)**

Contraction of involuntary muscle in the arterioles brings about a vasoconstriction. The presence of elastic fibres in blood arteries means that vasodilation occurs when the smooth muscle relaxes.

(b) bending at the knee joint to walk. **(2 marks)**

Skeletal muscles work in antagonistic pairs. When the quadriceps muscles contract the leg straightens. To bend the leg, the quadriceps muscles relax and the muscles on the back of the upper leg contract.

Now try this

1 All muscle tissue is rich in protein. Explain why. **(2 marks)**

2 Muscles typically work in antagonistic pairs, explain why. **(3 marks)**

Muscular contraction

Muscles contract through the interaction of two key proteins, actin and myosin, using ATP as an energy source.

Mechanism of muscular contraction

1 For contraction to occur, part of the myosin molecule (head group) attaches to a binding site on the actin filament; a cross-bridge is formed.

2 The head group bends, pulling the actin filament along, and ADP and Pi are released – the power stroke.

3 The attachment of a new ATP molecule to the myosin head breaks the cross-bridge.

4 The head group moves back to its original confirmation as ATP is hydrolysed to ADP and Pi, so another cross-bridge can be formed.

The power stroke

Contraction cycle continues if ATP is available and Ca²⁺ level in the sarcoplasm is high

The neuromuscular junction

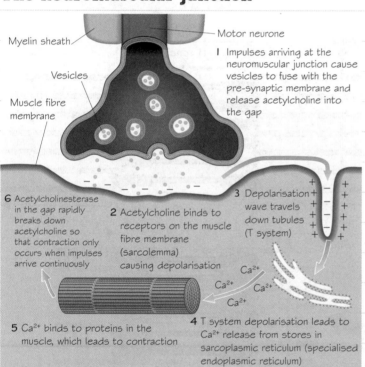

Myelin sheath

Motor neurone

Vesicles

Muscle fibre membrane

1 Impulses arriving at the neuromuscular junction cause vesicles to fuse with the pre-synaptic membrane and release acetylcholine into the gap

6 Acetylcholinesterase in the gap rapidly breaks down acetylcholine so that contraction only occurs when impulses arrive continuously

2 Acetylcholine binds to receptors on the muscle fibre membrane (sarcolemma) causing depolarisation

3 Depolarisation wave travels down tubules (T system)

Ca²⁺ Ca²⁺ Ca²⁺ Ca²⁺

5 Ca²⁺ binds to proteins in the muscle, which leads to contraction

4 T system depolarisation leads to Ca²⁺ release from stores in sarcoplasmic reticulum (specialised endoplasmic reticulum)

> The binding of Ca²⁺ to troponin causes the myosin binding sites on the actin filament (hidden by tropomyosin) to be revealed so the 'power stroke' can proceed.

Worked example

ATP is unstable and only small amounts of it exist in cells at any one time. Contraction of muscles uses lots of ATP and the ATP immediately available is used up after 1–2 seconds. Suggest how more ATP can be supplied to contracting muscles. **(3 marks)**

Aerobic respiration in muscle cell mitochondria produces ATP. Blood brings oxygen to muscle cells so glucose can be respired to generate ATP from the Krebs cycle and oxidative phosphorylation.

🧪 Practical skills — Monitoring muscle contraction

Investigations using the calf muscle removed from a frog's leg with the nerve still attached, show how an increase in stimulus frequency reduces time between contractions. Repeated large stimuli, close together, give a sustained and powerful contraction known as tetanus.

Single stimulus

Time→

Contraction recorded

Increasing level of stimulation

Repeated stimuli

Time→

Contraction recorded

Multiple stimuli lead to tetanus; eventually muscle fatigue reduces the level of contraction

Now try this

How would a lack of Ca²⁺ affect muscle contraction?

(2 marks)

Exam skills

This exam-style question uses knowledge and skills you have already revised. Your maths and practical skills are both tested here in the context of reaction time and the nervous system (look back at pages 112–115).

Worked example

Caffeine is a known stimulant. The table below shows the amount of caffeine present in various types of food and drink:

Food or drink	Serving size	Caffeine (mg)
Instant coffee	250 ml	120
Espresso coffee	50 ml	150
Cola drink	330 ml	35
Chocolate bar	55 g	10
Energy drink	330 ml	160
Black tea	250 ml	65
Green Tea	250 ml	50

(a) Which of the drinks contains the most caffeine per ml? Show your working. **(2 marks)**

espresso coffee: $\frac{150}{50} = 3\,mg/ml$

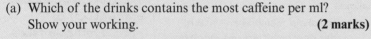

(b) According to the table, what is the percentage difference in caffeine content between a serving of green tea and a serving of instant coffee? Show your working. **(2 marks)**

$\frac{(120 - 50)}{50} \times 100 = 140\%$ more caffeine in the coffee

(c) A student decided they wanted to test the different drinks to see if a single serving affected a person's reaction rate. The student got four subjects to have one serving of the different drinks with breakfast over five consecutive days. The student found that there was no significant difference between the reaction times obtained for each drink.

Suggest two limitations in the student's investigation; explain how they might have affected the results and propose modifications to improve the method. **(6 marks)**

Limitation 1: Subjects may have consumed other drinks/foods for breakfast.
Effect: These could have contained substances which could have altered the results.
Modification: All subjects should have identical breakfasts.

Limitation 2: No times were specified after breakfast and before the test.
Effect: Varying times would mean that the caffeine had a longer or shorter time to take effect.
Modification: All subjects should have been instructed to have breakfast at exactly the same time.

Always make sure you show your working when directed. Marks can be gained by including the right numbers in calculations even if you make a mistake with the final calculation.

Maths skills Calculating the percentage difference between two values is a simple, but important skill. Make sure you start by working out the difference between the two values. You then need to read the question carefully to work out which number you divide the difference by. For instance, if you are asked about a percentage increase, then you divide by the lower value. Here you are simply asked for the percentage difference, so you could say the coffee has 140% more caffeine than the green tea per serving, or you could say the green tea has 58% less caffeine in it than the coffee (per serving) – both would be correct.

Practical skills This investigation highlights the importance of controlling variables. You are told that the student concluded that there was no difference in the effects of the caffeinated drinks. You could say that this conclusion lacked validity as the student failed to control certain variables.

Other limitations could include the age, mass and sex of the subjects. All of these factors would affect the way in which caffeine affected the subjects. The student did get the subjects to repeat the process 5 times, so repeating further in this instance might not produce more reliable results.

Photosynthesis and respiration

Photosynthesis involves the harvesting of light energy leading to the production of glucose and oxygen; these products can then be used by plant cells in aerobic respiration.

During the day, plants photosynthesise more than they respire. The extra sugars made can be stored as starch, or used to make cellulose, lipids or amino acids.

The **compensation point** is when the volume of oxygen produced by photosynthesis is equal to that used in aerobic respiration.

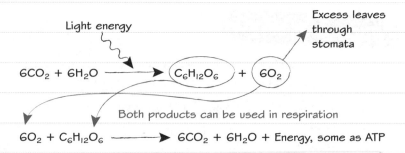

Light energy

Excess leaves through stomata

$$6CO_2 + 6H_2O \longrightarrow C_6H_{12}O_6 + 6O_2$$

Both products can be used in respiration

$$6O_2 + C_6H_{12}O_6 \longrightarrow 6CO_2 + 6H_2O + \text{Energy, some as ATP}$$

Photosynthesis produces glucose and oxygen. These can be used in respiration.

The structure of the chloroplast

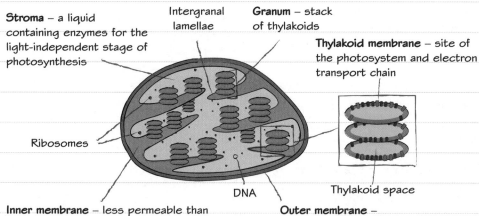

Stroma – a liquid containing enzymes for the light-independent stage of photosynthesis

Intergranal lamellae

Granum – stack of thylakoids

Thylakoid membrane – site of the photosystem and electron transport chain

Ribosomes

DNA

Thylakoid space

Inner membrane – less permeable than outer membrane, folded into thin plates (lamellae) and flattened sacs (thylakoids) in stacks (grana)

Outer membrane – permeable (to small ions)

A chloroplast and a granum. A chloroplast is a disc-shaped organelle, 2–10 μm in length, with a double membrane.

The LDS

The light-dependent stage (LDS) of photosynthesis takes place across the thylakoid membrane. The light-independent stage (LIS) occurs in the stroma.

Look at pages 122 and 123 for more details about these processes.

Worked example

An English chemist, Joseph Priestley put a sprig of mint into a transparent closed space with a lit candle. The candle quickly went out. After 27 days, he relit the extinguished candle and it burned perfectly well in the air that previously would not support combustion. Explain these observations.

(3 marks)

The candle went out because it had used the available oxygen during combustion. During the 27 days the mint photosynthesised and produced oxygen. The extinguished candle could be re-lit and use the oxygen as it burned.

Now try this

1. 'In the dark, plants behave like animals.' Suggest what this statement means in terms of photosynthesis and respiration.
   (3 marks)

2. Glucose is very soluble in water.
   (a) How would the production of glucose affect the water potential of a chloroplast? **(1 mark)**
   (b) What might happen to the chloroplast as a consequence and how might this be prevented? **(2 marks)**

Photosystems, pigments and thin layer chromatography

Chloroplasts contain light-harvesting pigments arranged in photosystems that capture light energy, enabling photosynthesis to proceed.

Photosystems

Sometimes called antennae complexes, photosystems are funnel-shaped, light-harvesting clusters of photosynthetic pigments held in place by protein complexes in the thylakoid membrane.

Photosynthetic pigments are molecules that absorb light energy. Each pigment absorbs certain wavelengths of light and reflects others.

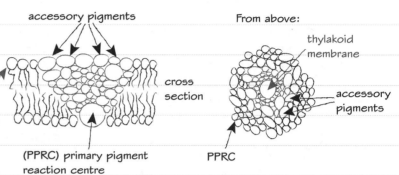

accessory pigments

thylakoid membrane

cross section

From above:

thylakoid membrane

accessory pigments

(PPRC) primary pigment reaction centre

PPRC

Different pigments

Chlorophyll a in the PPRC:

- There are two forms, P680 and P700 (numbers refer to absorption peaks).
- Both appear green-yellow.
- Absorbing light excites electrons previously associated with magnesium.

Chlorophyll b is an accessory pigment:

- It has absorption peaks at 500nm and 640nm.
- It appears blue-green.

Carotenoids:

- Include carotene and xanthophyll which are also accessory pigments.
- They appear orange-yellow.

Pigments can be identified by calculating the R_f value:

$$R_f = \frac{\text{distance moved by the compound (mm)}}{\text{distance moved by the solvent (mm)}}$$

A chromatogram of photosynthetic pigments extracted from spinach

🧪 Practical skills

Separating and identifying photosynthetic pigments

Using a suitable solvent such as propanone, photosynthetic pigments can be extracted from plant material by crushing it in a pestle and mortar. These pigments can then be separated using a process known as thin layer chromatography (TLC). (Look back at page 21 for more detail about TLC.)

Chloroplasts are the largest of all organelles (other than nuclei). In plant cells they move around by cytoplasmic streaming so they can absorb more light.

Worked example

Photosynthetic pigments in plants produce an absorption spectrum. Explain what is meant by the term 'absorption spectrum' and why most plants tend to be green in colour. **(3 marks)**

Plants have many different pigments so an absorption spectrum shows the range of light wavelengths absorbed by all the pigments. Plants appear green because they lack pigments that absorb green wavelengths of light.

Now try this

1 Some plants on the forest floor can have very dark green or even red leaves. Suggest why this is. **(2 marks)**

2 Where would you find accessory pigments? **(1 mark)**

Pigments absorb energy at a range of wavelengths. Try and learn the absorption peaks for different pigments rather than just the colours.

Light-dependent stage

In the light-dependent stage (LDS) of photosynthesis, light energy is used to drive the production of ATP and reduced NADP, which can be used as a source of energy for other metabolic processes.

Light-dependent stage

The LDS takes place across the thylakoid membranes of the chloroplast. Two photosystems (PS I and PS II) harvest light energy, converting it into a flow of electrons which ultimately drives the production of ATP and reduced NADP.

PS II contains an enzyme that splits water in the presence of light

replace electrons lost from chlorophyll (PS II)

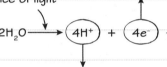

$$2H_2O \longrightarrow 4H^+ + 4e^- + O_2$$

used to establish H$^+$ gradient for ATP synthesis

diffused out through stomata

Cyclic photophosphorylation

This involves only PS I. Excited electrons pass through a chain of electron acceptors back to the chlorophyll molecule from which they were lost. NADP is not reduced but small amounts of ATP are made. Photolysis of water does not occur.

The small amounts of ATP made quickly in cyclic photophosphorylation, as well as being used in the LIS, can be used to establish a water potential gradient between guard cells and the surrounding tissues, which leads to the opening of stomata.

Photolysis of water on the thylakoid membrane

Non-cyclic photophosphorylation

- Energy is released as electrons pass between photosystems (PS I and PS II); some of that which is used to pump protons into the thylakoid space.

- The movement of electrons from a low to high energy state and back again is referred to as the Z-scheme.

- Electrons from PS I and protons (having passed through ATP synthase) are used by NADP reductase to reduce NADP.

- In cyclic photophosphorylation, electrons cycle back through the electron transport chain, returning to PS I rather than being used to reduce NADP.

Light can be thought of as travelling in particles (photons). When a photon hits a chlorophyll molecule its energy is transferred, 'exciting' electrons.

Key

〰〰 Light
→ Path of electrons
→ Path of hydrogen ions
● Electron carriers
◆ Site of photolysis of water

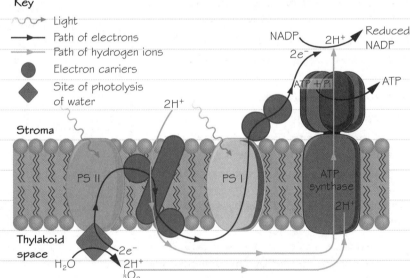

Worked example

Cyclic and non-cyclic photophosphorylation take place in chloroplasts of plant cells. Construct a table to show the differences between the two processes. **(4 marks)**

Non-cyclic	Cyclic
Involves PSI and PSII	Involves PSI only
Needs photolysis	No photolysis
Extensive transfer of electrons	Electrons flow back to original chlorophyll
Produces ATP and NADPH	Small amounts of ATP are made

Now try this

1 Explain why cyclic photophosphorylation would still proceed in the absence of NADP. **(2 marks)**

2 Why is it important that light promotes stomatal opening? **(2 marks)**

Light-independent stage

ATP and reduced NADP from the LDS are used directly in the light-independent stage (LIS) to fix carbon dioxide, which can then be used in the building of complex, organic molecules.

The Calvin cycle (LIS)

The enzyme RuBisCO (ribulose bisphosphate carboxylase-oxygenase) combines a 5-carbon molecule (ribulose bisphosphate, RuBP) with carbon dioxide (CO_2) to make an unstable 6-carbon intermediate. This breaks down to two 3-carbon molecules which can be used to synthesise sugars, amino acids and lipids.

The LIS uses ATP and reduced NADP from the LDS and CO_2, and it occurs in the stroma of the chloroplast.

CO_2 diffuses into the leaf through open stomata, passing through air spaces and into plant cells.

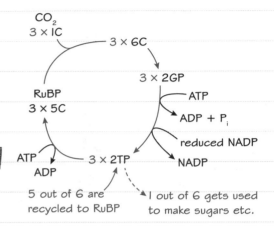

Synthesising new organic molecules

Triose phosphate (TP) is used in the synthesis of:

• carbohydrates

• lipids

• amino acids.

It is also recycled to regenerate RuBP.

Glycerate 3-phosphate (GP) is also used in the synthesis of amino and fatty acids.

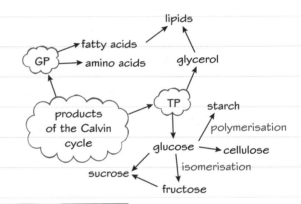

RuBisCO

RuBisCO is the most abundant enzyme on the planet. It evolved during a time when CO_2 concentrations were much higher, and O_2 levels were a lot lower. O_2 also fits the active site of RuBisCO and, above 25 °C, photorespiration rather than photosynthesis starts to occur, creating toxic waste products that require ATP to be broken down.

Worked example

You can use the abbreviations GP, TP and RuBP in your answers to save time in exams.

Complete the table to show the roles of key components in the LDS and the LIS of photosynthesis.   **(5 marks)**

Molecule	Role
ATP	Phosphorylates GP
Oxygen	Made as byproduct of the LDS
Triose phosphate (TP)	A 3-carbon molecule used in the synthesis of sugars
Water	Undergoes photolysis in the LDS
redNADP	A coenzyme oxidised in the LIS
RuBisCO	An enzyme that fixes CO_2

Now try this

1 Plants would not grow in an atmosphere of pure oxygen. Explain why.   **(2 marks)**

2 Outline the roles of ATP in the LIS.   **(2 marks)**

3 Explain why only small amounts of RuBP are found in the stroma.   **(2 marks)**

Factors affecting photosynthesis

Light intensity, carbon dioxide concentration and temperature all act as limiting factors in photosynthesis. Remember that limiting factors are factors that are present at the least or lowest favourable value for a reaction to proceed.

Light intensity and CO_2 concentration

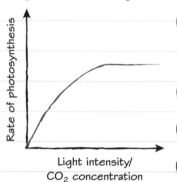

Rate of photosynthesis

Light intensity/ CO_2 concentration

1 Initially, as light intensity / CO_2 concentration increases, so does the rate of photosynthesis.

2 At higher intensities, the rate plateaus.

3 Some other factor is now said to be limiting.

4 Increasing that factor will result in a further increase in rate.

Temperature

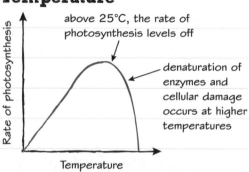

Rate of photosynthesis

Temperature

above 25°C, the rate of photosynthesis levels off

denaturation of enzymes and cellular damage occurs at higher temperatures

The LDS is not affected greatly by temperature fluctuations, but the LIS is a series of enzyme-controlled reactions and as such the rate approximately doubles for each 10°C rise. Above 25°C, however, photorespiration begins to occur.

The effect on levels of GP, RuBP and TP

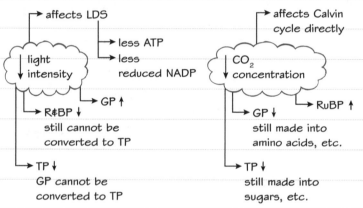

light intensity
→ affects LDS
→ less ATP
→ less reduced NADP
→ GP ↑
→ R&BP ↓ still cannot be converted to TP
→ TP ↓ GP cannot be converted to TP

CO_2 concentration
→ affects Calvin cycle directly
→ GP ↓ still made into amino acids, etc.
→ RuBP ↑
→ TP ↓ still made into sugars, etc.

Worked example

Water stress affects the rate of the processes involved in photosynthesis. A lack of water can lead to stomatal closure. Explain why the rate of photosynthesis might be affected by a lack of water.

(3 marks)

The closing of stomata would reduce gas exchange so less CO_2 might be available to combine with RuBP. The rate of transpiration would also be reduced. This would affect the turgidity of cells, leading to a plant wilting, and also reducing the rate at which minerals such as nitrates were brought to the leaf, reducing the rate of amino acid production.

Less water would be available to act as a reagent in the LDS of photosynthesis.

Investigating the rate of photosynthesis

Practical skills

Bung
Boiling tube
Elodea
In light

Hydrogen-carbonate indicator solution

In dark

Using an indicator in solution to measure the uptake of CO_2:

Hydrogen carbonate indicator solution is red when neutral, yellow at pH 6 and purple at a pH slightly above 7.

As CO_2 is used the solution will turn increasingly purple. This can be measured using a colorimeter. The rate at which colour change occurs can be linked to light intensity by varying the distance of a light source from the solution.

Light intensity can be calculated using the formula: $\frac{1}{d^2}$ (where d is the distance from the light source).

Now try this

1 CO_2 can still be fixed for some time at night. Explain how. **(2 marks)**
2 RuBP levels fall as light intensity drops. Explain why. **(2 marks)**

The need for cellular respiration

Cellular respiration is the process whereby energy stored in complex organic molecules is used to make ATP.

ATP uses

Animals, plants and microorganisms must all respire to generate ATP. ATP can be used for a variety of cellular functions:

- secretion (exocytosis) – change in cytoskeleton
- active transport – moving substances against concentration gradients
- movement – cilia, flagella (motor proteins)
- endocytosis – change in cytoskeleton
- activation – phosphorylating other molecules
- replication – DNA
- synthesis – proteins, lipids, for example.

Energy flow

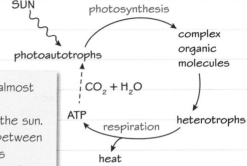

The energy for almost all life on earth originates from the sun. Carbon cycles between photoautotrophs (plants), heterotrophs and the atmosphere.

ATP structure and importance

ATP is a phosphorylated nucleotide. It is the hydrolysis of ATP to ADP and Pi that makes energy available in small, manageable amounts (look back at page 24).

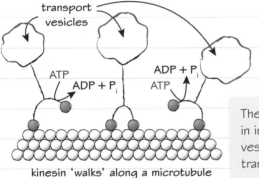

kinesin 'walks' along a microtubule

The role of ATP in intracellular vesicle transportation

The role of ATP in active transport

Active transport uses ATP to move substances against their gradients across membranes.

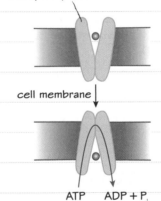

The hydrolysis of ATP allows a transport protein to change shape, moving a substance against a concentration gradient across a cell membrane.

Worked example

Your metabolism is the sum total of all the chemical reactions that take place in the body. It can be split into **anabolic** reactions which involve building large molecules from smaller molecules and **catabolic** reactions which involve breaking down large molecules into smaller ones.

Categorise the following metabolic processes: **(5 marks)**

Process	Anabolic or catabolic
Synthesis of actin filaments	Anabolic
Formation of insulin	Anabolic
Digestion of starch	Catabolic
Conversion of glucose to glycogen	Anabolic
Lysozyme activity in a phagolysosome	Catabolic

Energy states

Energy can be considered to occur in different states and the body contains transducers that convert energy between different states; for example, light energy to nervous impulse (photoreceptors in the eye).

Energy is released, not 'made'. It can be lost from cells in the form of heat – endotherms harness this heat to maintain body temperature.

Now try this

1 Why is ATP a particularly useful source of energy? **(2 marks)**
2 What are the proteins called that enable vesicle movement in cells? **(1 mark)**

Glycolysis

Glycolysis is a biochemical pathway occurring in the cytoplasm, common to all forms of cellular life which converts glucose to pyruvate, generating small quantities of ATP and reduced NAD.

Glycolysis stage 1

At this stage there has been a net usage of ATP as phosphate groups are added to the 6-carbon sugar, glucose.

glucose

↓ ┌ ATP
　└→ ADP

glucose 6-phosphate

│ isomerisation
↓

fructose 6-phosphate

↓ ┌ ATP
　└→ ADP

hexose 1,6-bisphosphate

2 × ATP used.
'energy investment' stage

high energy molecule

The process of glycolysis

The phosphorylation of glucose to hexose bisphosphate

Glycolysis stage 2

Hexose bisphosphate is spilt to produce two triose phosphate molecules.

hexose 1,6-bisphosphate

↓

2 × triose phosphate (TP)

the rest of glycolysis proceeds in duplicate from now on

the same molecule that occurs as an intermediate in the Calvin cycle

Splitting of hexose bisphosphate

Glycolysis stage 3

Dehydrogenase enzymes remove two hydrogen atoms from each triose phosphate making two molecules of reduced NAD.

2 × triose phosphate

substrate-level phosphorylation

↓ NAD × 2

2 × ATP → redNAD × 2

2 × intermediate compound

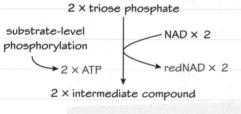

Oxidation of triose phosphate

Glycolysis stage 4

2 × intermediate compound

↓

substrate-level phosphorylation

↘ 2 × ATP

2 × pyruvate

link reaction in the matrix of the mitichondrium

products of glycolysis
2 × ATP(net)
2 × redNAD
2 × pyruvate

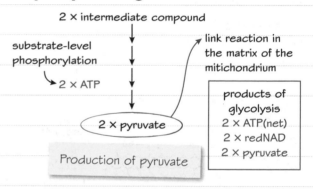

Production of pyruvate

Worked example

Nearly all living things use the glycolytic pathway as a source of ATP. Suggest why further pathways evolved given that ATP can be made in this process.　　**(3 marks)**

Glycolysis only produces two molecules of ATP per glucose molecule. A lot of energy is left in the pyruvate molecule. Multicellular organisms have high energy demands and they require a lot more ATP to be made to fulfil these demands, for example muscle contraction.

Glycolysis is anaerobic

Glycolysis does not require oxygen to proceed. As long as NAD is available, it can continue. The production of ethanol in yeast and lactic acid in mammals both ensure NAD is available for glycolysis.

'Suggest' style questions require you to use the information given AND your own knowledge to put forward an idea that answers the question; use examples if you can to develop your ideas.

Now try this

1　Glycolysis produces a net yield of 2 ATP. Explain what this means.　　**(2 marks)**
2　Redox reactions are important in glycolysis. Give an example of a molecule that becomes reduced and one that becomes oxidised.　　**(2 marks)**
3　Where does glycolysis occur?　　**(1 mark)**

Structure of the mitochondrion

Mitochondria are organelles found in eukaryotic cells. They are the sites of the aerobic stages of respiration: the link reaction, Krebs cycle and oxidative phosphorylation.

Structure of a mitochondrion

Shapes vary, but most are 0.5–1.0 μm in diameter and 2–5 μm long.

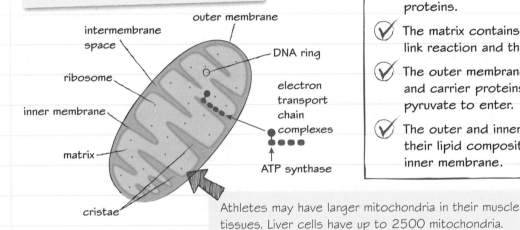

Relating structures to functions

✓ The DNA and ribosomes enable mitochondria to make some of their own proteins.

✓ The matrix contains the enzymes for the link reaction and the Krebs cycle.

✓ The outer membrane contains channel and carrier proteins, some of which allow pyruvate to enter.

✓ The outer and inner membranes differ in their lipid composition. See below for the inner membrane.

Athletes may have larger mitochondria in their muscle tissues. Liver cells have up to 2500 mitochondria.

The inner membrane

- folded into cristae to give a large surface area
- largely impermeable to smaller ions, including protons (H^+)
- embedded with electron carriers and ATP synthase enzymes
- electron transport protein complexes pass electrons between each other, with associated Fe atoms becoming alternately reduced (Fe^{2+}) then oxidised (Fe^{3+})
- some complexes have proton pumps which move protons into the intermembrane space
- ATP synthase makes ATP as protons flow through it, down a proton gradient.

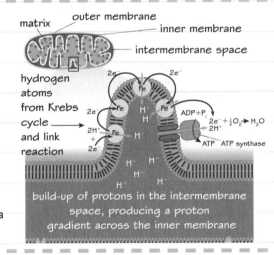

build-up of protons in the intermembrane space, producing a proton gradient across the inner membrane

Worked example

Explain the significance of the relative impermeability of the inner membrane of the mitochondrion. **(3 marks)**

It ensures a proton gradient can be established. The inner membrane is impermeable to protons. When they are pumped into the intermembrane space, the only way they can move back across it is via ATP synthase, which ensures ATP production.

When you learn the structure of the mitochondrion try to always link it to a particular function – what is it about that structure that enables it to perform its function effectively?

Now try this

1 Where would you find oxaloacetate, a component of the Krebs cycle? **(1 mark)**

2 Why are the cristae important? **(2 marks)**

3 What features support the idea that mitochondria are derived from prokaryotes? **(2 marks)**

Link reaction and the Krebs cycle

Both the link reaction and the Krebs cycle take place in the mitochondrial matrix.

The link reaction and Krebs cycle

Krebs cycle: Each acetate group combines with a molecule of oxaloacetate to make citrate. The cycle proceeds through to the reconversion of citrate to oxaloacetate with the production of three molecules of reduced NAD, one molecule of reduced FAD (reduction), two molecules of CO_2 (decarboxylation) and one molecule of ATP (substrate-level phosphorylation). The CO_2 will diffuse out of the mitochondria as it is produced.

The link reaction: Two molecules of pyruvate are decarboxylated, oxidised and combined with CoA to produce two molecules of CO_2, two molecules of reduced NAD and two molecules of acetyl coA.

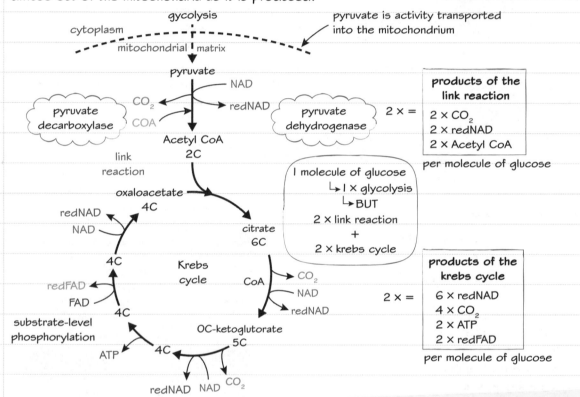

products of the link reaction		
$2 \times CO_2$		
$2 \times$ redNAD		
$2 \times$ Acetyl CoA		

per molecule of glucose

I molecule of glucose
↳ I × glycolysis
↳ BUT
2 × link reaction
+
2 × krebs cycle

products of the krebs cycle		
$6 \times$ redNAD		
$4 \times CO_2$		
$2 \times$ ATP		
$2 \times$ redFAD		

per molecule of glucose

The importance of coenzymes:

NAD and FAD are coenzymes for dehydrogenase enzymes. They play a key role in every stage of aerobic respiration.

Coenzyme A carries acetate molecules made during the link reaction on to the Krebs cycle.

It is important to keep track of the hydrogen atoms throughout each stage of respiration. Coenzymes, NAD and FAD are involved in the oxidation of respiratory intermediates, becoming reduced themselves. The hydrogen they pick up is used later in respiration to establish a proton gradient across the inner mitochondrial membrane.

Worked example

Including glycolysis, the link reaction and the Krebs cycle, make a list of all the ATP, reduced NAD, reduced FAD and carbon dioxide produced from the respiration of a single glucose molecule. **(4 marks)**

Substance	Glycolysis	Link reaction	Krebs cycle	Total so far
ATP	2	O	2	4
reduced NAD	2	2	6	10
reduced FAD	O	O	2	2
CO_2	O	2	4	6

Now try this

1 Explain why only small amounts of oxaloacetate are made in the matrix. **(1 mark)**

2 The mitochondrial membranes are not permeable to reduced NAD. Explain what might happen if they were. **(3 marks)**

3 Tryptophan is an amino acid used in the synthesis of NAD. What might happen if you had a diet that lacked tryptophan?
(3 marks)

Oxidative phosphorylation

Oxidative phosphorylation is the production of ATP in the presence of oxygen via the process of chemiosmosis. This involves the electron transport chain (ETC), proton gradients and ATP synthase. These are the final stages of aerobic respiration.

Chemiosmosis leading to oxidative phosphorylation

Chemiosmosis is the flow of hydrogen ions across a phospholipid membrane via the transmembrane portion of ATP synthase.

> The complexes of the ETC and ATP synthase are found on the mitochondrial cristae.

1 Reduced NAD is re-oxidised by NADH dehydrogenase.

2 The hydrogen removed is split into H^+ and e^-.

3 The e^- are passed along electron carriers in the ETC.

4 The energy released from the electrons is used to pump H^+ across the inner mitochondrial membrane.

5 FAD also contributes electrons, but the H^+ are not pumped directly through the membrane.

6 Cytochrome oxidase enables oxygen (O_2) to accept two electrons at the end of the ETC, combining them with $2H^+$ from the matrix to make water (H_2O).

$$O_2 + 4e^- + 4H^+ \xrightarrow{\text{cytochrome oxidase}} H_2O$$

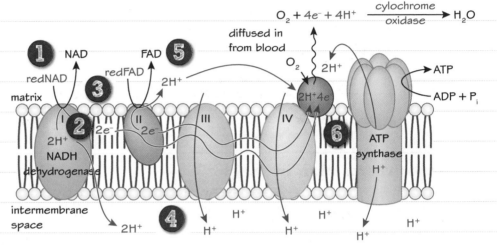

ATP synthase and the evidence for chemiosmosis

There is a difference in pH between the matrix and the intermembrane space due to the pumping of protons by complexes in the ETC. This piece of evidence helped to develop the chemiosmotic theory.

ATP synthase is the enzyme responsible for producing large quantities of ATP in oxidative phosphorylation.

> Early electron microscopy showed ATP synthase as mushroom-shaped particles.

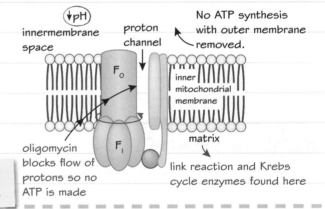

Worked example

Cytochrome oxidase is inhibited in a non-competitive way by cyanide. What does this mean?
What are the consequences as far as the ETC is concerned?

(5 marks)

Cyanide binds to an allosteric site (not the active site) on the enzyme causing the active site to change shape so enzyme–substrate complexes cannot form.

This may be permanent. If cytochrome oxidase doesn't function then electrons are no longer accepted by oxygen and they cease to flow along the ETC. Reduced NAD will no longer be reoxidised and the aerobic parts of respiration will come to a halt. The amount of ATP produced will decrease.

> Cyanide is a respiratory toxin, with a single molecule effectively disabling the whole ETC.

Now try this

1 How many complexes pump protons across the inner membrane in the ETC?   **(1 mark)**

2 Describe the pathway taken by oxygen from the blood to the matrix of the mitochondria.

(3 marks)

Anaerobic respiration in eukaryotes

Practical skills When oxygen is lacking, the ETC cannot function so the Krebs cycle and link reaction also stop. The reduced NAD generated during the oxidisation of glucose must be reoxidised so that glycolysis can continue.

The lactate pathway in mammals

Lactate fermentation is the pathway used in mammalian cells to reoxidise NAD, so that the glycolytic pathway can function. It occurs during times of anaerobic respiration, for example intense exercise, leading to the production of a small amount of ATP, so that important metabolic processes can continue.

Lactate is transported to the liver via the blood. When more oxygen is available it can be converted back to pyruvate (remember the Krebs cycle, on page 128).

Ethanol fermentation in yeast

Fungi such as yeast use ethanol formation to reoxidise NAD. It is a two-step process, which sees a decarboxylation step followed by a reduction step.

Ethanol (and CO_2) are lost from the yeast cell.

In both cases, NAD is made available to accept hydrogen from triose phosphate in glycolysis.

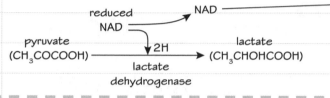

pyruvate ($CH_3COCOOH$) → [lactate dehydrogenase, 2H] → lactate ($CH_3CHOHCOOH$)
reduced NAD → NAD

pyruvate ($CH_3COCOOH$) → [pyruvate decarboxylase] CH_3CHO ethanal + CO_2 → [ethanal dehydrogenase, 2H] → CH_3CH_2OH ethanol
reduced NAD → NAD

Aerobic respiration in yeast

This simple set-up allows any CO_2 produced to be absorbed by the potassium hydroxide so the volume of oxygen used for respiration can be directly measured.

Vol. = distance moved by water

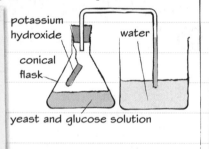

potassium hydroxide
water
conical flask
yeast and glucose solution

Anaerobic respiration in yeast

The layer of oil restricts the yeast's access to oxygen and respiration becomes anaerobic. The limewater will turn milky as CO_2 bubbles into it.

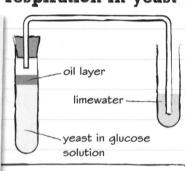

oil layer
limewater
yeast in glucose solution

Worked example

Complete the table comparing anaerobic respiration in yeast and mammals. **(5 marks)**

	Yeast	Mammals
Enzymes involved	pyruvate decarboxylase and ethanal dehydrogenase	lactate dehydrogenase
Hydrogen acceptor	ethanal	pyruvate
NAD reoxidised?	yes	yes
CO_2 produced?	yes	no
End products	ethanol and carbon dioxide	lactate

Note the names of the enzymes; they are dehydrogenase enzymes but in these instances they are reducing the substrate and oxidising the coenzyme.

In yeast the rate of ATP production during anaerobic glycolysis can be up to 100 times faster than oxidative phosphorylation. This uses a lot of glucose very quickly and creates ethanol, which can ultimately kill yeast if it builds up above 15%. Yeast are described as facultative anaerobes because of this – they can survive without oxygen as long as enough glucose is present.

Now try this

1 Covering bread dough can make it rise more quickly. Explain why. **(3 marks)**

2 What is the role of the potassium hydroxide in the above experiment? **(1 mark)**

Energy values of different respiratory substrates

Respiratory substrates are organic substrates that can be used for respiration; different substances yield different amounts of energy per unit mass.

Carbohydrates as respiratory substrates

As the majority of ATP is made by ATP synthase powered by the flow of H^+ ions, it follows that a substrate that contains a high proportion of hydrogen will lead to more ATP being produced.

Glucose is the main respiratory substrate.

- Animals store glucose as glycogen.
- Plants store glucose as starch.

As carbohydrates contain substantial amounts of oxygen in addition to carbon and hydrogen, they are converted to ATP less efficiently than some other respiratory substrates.

Mean energy value = $15.8 \, kJ \, g^{-1}$

Proteins as respiratory substrates

Proteins are respired during times of starvation, fasting and exercise. The number of hydrogen atoms per molecule is slightly higher than for carbohydrates, so more ATP can be produced from the same mass via oxidative phosphorylation.

Amino acids can be deaminated and then respired with the hydrocarbon portions entering at various stages of the aerobic respiration pathway.

Mean energy value = $17.0 \, kJ \, g^{-1}$

Fats as respiratory substrates

Fatty acids (look back at page 12 as a reminder) are long hydrocarbon chains joined to a glycerol molecule. They therefore have more C–H bonds and H atoms than carbohydrates or proteins. Glycerol can be converted to triose phosphate and fed into glycolysis.

- β-oxidation pathway in the matrix of the mitochondria produces 2-carbon acetyl groups, reduced NAD and reduced FAD.
- Acetyl groups are fed in at the acetyl CoA stage.
- Large numbers of H^+ are delivered to the ETC by reduced NAD and reduced FAD.

Mean energy value = $39.4 \, kJ \, g^{-1}$

 Maths skills **Respiratory quotient**

$$RQ = \frac{\text{Volume of } CO_2 \text{ evolved}}{\text{Volume of } O_2 \text{ absorbed}}$$

It takes more oxygen to respire proteins and fats, than carbohydrates. The relative volumes of O_2 used and CO_2 evolved can be used to produce a respiratory quotient that allows certain conclusions to be made about the respiratory substrates being used by an organism. A quotient of around 1.0 suggests a diet of mainly carbohydrates; 0.8 and 0.7 infer more proteins or fats are being respired respectively.

Worked example

A respirometer was used to establish that a locust respiring aerobically used $0.7 \, cm^3$ of O_2 in an hour, whilst producing $0.5 \, cm^3$ of CO_2.

Calculate the respiratory quotient. Suggest what this indicates about the diet of the locust. **(2 marks)**

$$\frac{0.5}{0.7} = 0.71$$

This suggests that the diet of the locust had been primarily fat based.

A respirometer is shown on page 132.

Now try this

1 Where in the cell does the oxidation of fatty acids take place? **(1 mark)**

2 For an animal undergoing a period of starvation, explain why a respiratory quotient of 1.0 would be unlikely. **(3 marks)**

Factors affecting respiration

🧪 **Practical skills** You can use a respirometer to investigate factors affecting respiration.

Measuring oxygen consumption

A respirometer measures the rate of consumption of oxygen by a living organism. When a substance that absorbs the CO_2 produced by respiration is present then the total volume of gas within an enclosed space (such as a conical flask) will decrease by the volume of oxygen being used. A rate can be calculated by determining the volume used and dividing by the time taken. By changing food sources (respiratory substrates), temperature and other variables, the rate of oxygen consumption can be used to indicate the level of respiration taking place.

> When animals are used in this type of experiment, ethical considerations must be made as to the welfare of the animal.

Measuring the volume of oxygen used over a period of time.

The potassium hydroxide solution absorbs the CO_2 released by the respiring seeds. This allows the volume of oxygen used for respiration to be determined.

It is important that the potassium hydroxide solution plus water in A, equals the total volume of the seeds and solution in B. This is because any changes in temperature will affect the pressure in the vessels and thus the gas volumes. By ensuring the same total volume in each case, any change in temperature will not affect the validity of the results.

The syringe allows the system to be reset after each experiment.

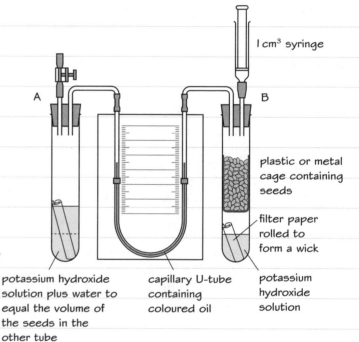

1 cm³ syringe

plastic or metal cage containing seeds

filter paper rolled to form a wick

potassium hydroxide solution plus water to equal the volume of the seeds in the other tube

capillary U-tube containing coloured oil

potassium hydroxide solution

> Make sure you understand how this equipment works and that you can answer questions which involve using it in an investigation.

Worked example

Describe how you would use a respirometer to investigate how temperature affects respiration rates in germinating seeds. **(6 marks)**

The apparatus could be placed into a thermostatically controlled water bath and left for 20 minutes, after which time the distance the manometer fluid moved would be measured and the volume uptake of oxygen per minute would be calculated. The syringe would be used to reset the manometer fluid before repeating two more times.

A mean would be determined from the three results. The experiment could be repeated at different temperatures (20°C to 60°C at 5° intervals). The potassium hydroxide would absorb the CO_2 allowing the O_2 evolved to be measured. The same volume and species of seed should be used for each temperature.

> Be precise about times and what you would measure in your answer. Also think about how the data would be processed, the need for repeats and what variables would need to be controlled.

Now try this

1 If you used invertebrates rather than seeds, would your methods need to change? **(2 marks)**

2 Why is the potassium hydroxide necessary in this investigation? **(1 mark)**

Exam skills

This exam-style question uses knowledge and skills you have already revised. Your knowledge and practical skills are both tested here in the context of plant growth and photosynthesis (look back at pages 108 and 122).

Worked example

(a) A pea plant growing from seed will climb up a supportive trellis. Following germination, describe and explain the various growth responses taking place in the plant. **(4 marks)**

Roots are positively geotropic so they grow down to anchor the plant and access water. Shoots are positively phototropic so they grow upwards, towards the light. Tendrils are positively thigmotropic so they respond to touch and grow around the trellis.

(b) DCPIP is a dye that changes from navy to colourless when it becomes reduced. It can be used to determine how rapidly the light-dependent stage (LDS) of photosynthesis is occurring as it is preferentially reduced over NADP. A chloroplast suspension can be made using blended spinach leaves and a 2% sucrose solution. Plan an experiment to investigate how different wavelengths of light affect the LDS of photosynthesis. **(6 marks)**

1. Split the chloroplast suspension into two beakers – one will be the control.

2. Add enough DCPIP to one of the beakers to produce a distinguishable blue colour.

3. Cover the DCPIP chloroplast mixture with foil to prevent ambient light causing reduction.

4. Take two capillary tubes; fill one with the control suspension and one with the DCPIP chloroplast mixture.

5. Immediately place them both under red light. Record how long it takes for the DCPIP chloroplast mixture to return to the colour of the control.

6. Repeat with other coloured lights.

(c) The chloroplast suspension was spun in a centrifuge until a pellet formed. The graph opposite shows how the absorbance of light changed over time when DCPIP was added to a tube containing the resuspended pellet and a tube containing the original supernatant, and both tubes were exposed to a bright light. Determine the gradient for both tubes between 3 and 6 minutes of light exposure.

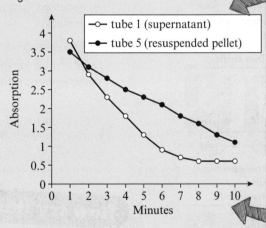

Supernatant:

At 3 minutes the absorption is 2.3.

At 6 minutes the absorption is 0.9.

$0.9 - 2.3 = -1.4$

$\dfrac{-1.4}{3} = -0.47$

Resuspended pellet:

At 3 minutes the absorption is 2.8.

At 6 minutes the absorption is 2.1.

$2.1 - 2.8 = -0.7$

$\dfrac{-0.7}{3} = -0.23$

This answer is structured well. It is split into three parts, is easy to read and descriptions are followed by an explanation. Note the use of the term 'positively' – it is important to qualify the growth response in this way.

Do not be put off if organisms or chemicals you have never heard of come up in an exam question. Look for similarities and links to things you do know about. Here the chemical DCPIP is introduced. You are told that it gets reduced in place of NADP in the LDS of photosynthesis. Use this information to get started.

 Practical skills You could improve this answer by using volumes and units in your plan and talking about variables that would need to be controlled, repeating the process and how you would process the data.

 Practical skills Centrifuges are used to separate substances by spinning them at very high speeds. The key information here is actually in the graph.

 Maths skills To calculate gradient $= \dfrac{\text{change in } y}{\text{change in } x}$

Place a ruler horizontally against the plotted point for 3 minutes for the supernatant and read across to the y-axis. Do the same for 6 minutes. Take the value for 3 minutes from the value for 6 minutes. This gives you a minus number to indicate the gradient is sloping downwards. This is your 'change in y'.

Now, the change in x is easy to determine: simply the difference between 3 and 6, which is 3.

Divide the change in y by the change in x and you have your gradient. Now do this for the resuspended pellet line.

Gene mutation

A gene mutation is a mistake in the DNA sequence of a gene.

What causes gene mutations?

Gene mutations are caused by mistakes during DNA replication as well as a number of factors in the environment, including:

- UV light
- pollutants
- chemicals in cigarette smoke
- radiation
- oxidants.

These factors are called **mutagens**. They cause changes to the bases in the DNA, so that the bases are deleted, inserted or substituted.

Most mutations are repaired by proteins inside the nucleus. If a mutation is not repaired, it can lead to changes in the **structure** of proteins.

Effects on proteins

Point mutations in the genetic sequence change the triplet **codon** (see page 27 for the genetic code). Each **codon** codes for a particular amino acid.

- As most amino acids have two or more codons, the mutation can often be **neutral**.
- Insertion and deletion mutations are also known as **frameshift mutations** because they alter the point from where the codon is read. These are more likely to alter the amino acid sequence and so change the way in which the protein folds.
- If the new folding improves the function of the protein, it often improves the characteristic of that protein. This is a **beneficial** mutation.
- Often the change in protein folding results in a non-functioning or truncated (smaller) protein. This is a **harmful** mutation.

Changes in amino acid sequence

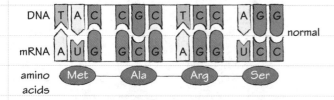

normal

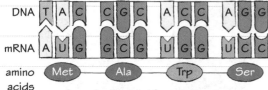

one base is changed for another base, changes only one amino acid

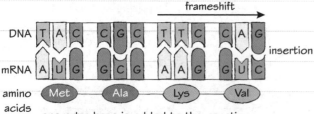

insertion

one extra base is added to the genetic sequence, causing a frameshift

deletion

one base is removed from the genetic sequence, causing a frameshift

Worked example

Explain how a change in the base sequence of a gene leads to a change in the tertiary structure of a protein. **(3 marks)**

> The tertiary of a protein is its 3D shape (look back at page 15 for a reminder about protein structure). This is controlled by bonds between the amino acids of the protein.

A mutation can be a substitution, insertion or deletion.

Triplet codon is changed.

Different amino acids can be added to the polypeptide sequence.

> A stop codon can also be introduced too soon, truncating the polypeptide chain, and leading to a less functional, or non-functional protein.

Now try this

1 Explain what type of mutation it would be if GAA were altered to GAG (you can look back at the genetic code on page 27). **(2 marks)**

2 Why are insertion and deletion mutations called frameshift mutations? **(3 marks)**

Gene control

The *lac* operon regulates the expression of lactose-specific genes in bacteria, and was the first example to be discovered. Gene expression can be regulated in different organisms in different ways.

The *lac* operon

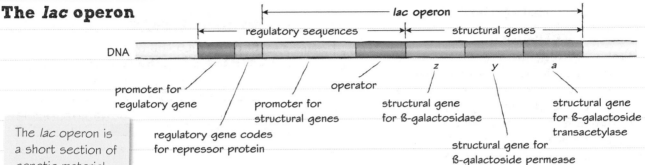

lac operon

regulatory sequences — structural genes

DNA

promoter for regulatory gene

regulatory gene codes for repressor protein

promoter for structural genes

operator

structural gene for ß-galactosidase

z   y   a

structural gene for ß-galactoside permease

structural gene for ß-galactoside transacetylase

> The *lac* operon is a short section of genetic material consisting of three genes together with a promoter, and an operator sequence.

In order to **transcribe** the genes, RNA polymerase must bind to the **promoter** and move over the **operator**. However, in the absence of lactose, the operator is blocked by a **repressor** protein.

When **lactose** is present, it binds to the repressor protein, changing the shape of the protein. The repressor protein can no longer bind to the operator and the genes can now be transcribed. The three genes produce proteins that are involved in the transport and hydrolysis of lactose and are only needed in the presence of lactose. This mechanism prevents the unnecessary production of proteins in the absence of lactose.

Gene control in eukaryotes

The transcription of eukaryotic genes is often controlled by **transcription factors**. These are proteins found in the nucleus that bind to the DNA and allow transcription to take place.

Some proteins are only activated after they have been bound by another molecule. For example, **protein kinase** is only activated after binding to a **cyclic AMP** (cAMP). Protein kinase is made of four subunits — two inhibitory subunits and two active subunits. cAMP binds to the inhibitory subunits, which break away from the active subunits, allowing them to carry out their function.

exon | intron | exon | intron | exon primary RNA

introns removed (spliced out) aided by enzymes containing RNA

splicing

exon   exon   exon

introns removed by spliceosome (complex of RNA and protein)

exon | exon | exon mature mRNA

> In eukaryotes, genes contain sections of **non-coding dna** called introns. The coding sections of the gene are called **exons**. **Post-transcriptional editing** removes the introns to produce mature mRNA. This is called **splicing**.

Worked example

Describe what is meant by post-transcriptional editing.   **(3 marks)**

Introns are sections of non-coding DNA (they are spliced out from primary mRNA) and exons are joined together.

> Non-coding DNA is part of the genetic sequence that does not code for the protein.
>
> Leaving the introns in the mRNA would mean that it would take longer for the ribosomes to translate the mRNA.

Now try this

1 How are genes controlled by proteins?   **(3 marks)**
2 Suggest what could happen if the repressor protein gene was mutated.   **(3 marks)**

Homeobox genes

Homeobox genes are very important. They control the development of the body plan in all multicellular organisms.

The role of homeobox genes in humans

Homeobox (Hox) genes were first identified in *Drosophila* (fruit flies), but were found to be present in all multicellular organisms. Each Hox gene codes for a protein called a **transcription factor**. Each gene contains a **homeobox sequence** that codes for a DNA-binding domain called the **homeodomain**. Transcription factors bind to DNA and control the expression of many other genes during development, by switching genes on and off. This controls how the body plan will develop.

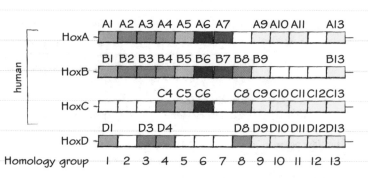

developing fetus

Hox genes are arranged into **clusters**. Most organisms have one cluster, but all mammals have four. Groups of genes that occupy the same position in different clusters (e.g. AI, BI and DI) are called **paralogues**. Humans and other mammals have 13 paralogues.

The coloured sections on the developing fetus correspond to the paralogous group of Hox genes that controls its development.

Homeobox sequences in other species

Arthropods (insects)

Echinoderms (starfish)

Tetrapod vertibrates

Amphibians
Reptiles
Birds
Mammals

> Homeobox sequences are highly **conserved** (very similar) across all species, including animals, plants and fungi. The same group of Hox genes is found in the same position on the Hox cluster. This means that Hox genes are very important for the development of all body plans, and that these genes must have originated from a common ancestor.

Now try this

1 Explain why there has been very little change by mutation in homeobox genes.   **(2 marks)**
2 How do homeobox genes control the development of body plans?   **(3 marks)**

The homeobox sequence is a 180-base-pair sequence found within Hox genes.

If the homeodomain part of the transcription factor binds to a promoter region on the DNA, it can repress or activate transcription of genes.

Mitosis and apoptosis

The development of the body form is controlled by two processes – mitosis (cell division) and apoptosis (programmed cell death).

Mitosis

The **zygote** (fertilised egg) is only one cell. A fully developed fetus has trillions of cells. The zygote divides by a type of cell division called **mitosis** (which you will remember from page 42). This is the division of body (somatic) cells. Each daughter cell is genetically identical to the parent cell.

Apoptosis

Apoptosis is programmed cell death. Under certain circumstances, the body causes some cells to die.

During fetal development, some cells die off to create the shape of the limbs.

For example, the hand bud forms as a round, smooth shape. Cells die off between where the fingers will be in order to make the shape of the fingers and thumbs.

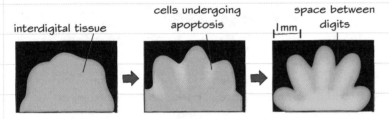

interdigital tissue · cells undergoing apoptosis · |1mm| · space between digits

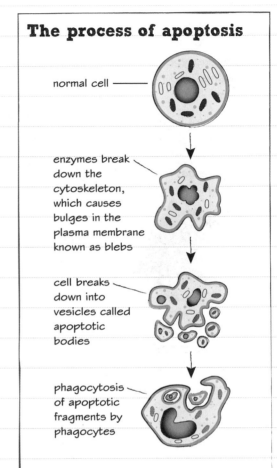

The process of apoptosis

normal cell

enzymes break down the cytoskeleton, which causes bulges in the plasma membrane known as blebs

cell breaks down into vesicles called apoptotic bodies

phagocytosis of apoptotic fragments by phagocytes

External and internal cell stimuli

The body cells can respond to stimuli from inside and outside of the body, for example stress. These stimuli can affect the genes that control the cell cycle. The stimuli cause cells to release signalling molecules, such as hormones, cytokines or growth factors. These molecules cause the cell cycle control genes to initiate apoptosis and kill off some of the body's cells. This protects the body from damage, for example, apoptosis could kill off cancerous cells.

Worked example

What is the mechanism of apoptosis?   **(3 marks)**

Cell begins to bleb.

Cells break into smaller fragments.

Fragments engulfed by phagocytosis.

Blebbing is the breakdown of the nuclear envelope and other cell membranes.

Blebbing is initiated by the genes that control the cell cycle.

Phagocytosis is carried out by macrophages.

Now try this

1 Explain the benefit of apoptosis in fetal development.   **(2 marks)**
2 How does a stimulus, such as viral infection, cause apoptosis?   **(3 marks)**

Variation

Variation is influenced by genes and the environment.

Genetic influence

All **phenotypes** (observable characteristics) have some genetic influence.

Some phenotypes are controlled by a single gene (**monogenic**) and do not have any environmental influence.

In mice, grey fur colour is dominant and white albino fur colour is recessive. The fur colour of the mice belongs to only one category. This is a type of **discontinuous** variation (see page 86).

Environmental influence

Other phenotypes have a strong environmental influence. These phenotypes tend to be controlled by several genes (**polygenic**).

For example, the stems of shaded plants will grow very long (**etiolation**). Each stem can be a range of different lengths. This is a type of **continuous** variation.

The stems of shaded plants will also appear yellow due to a loss of their chlorophyll, known as **chlorosis**.

Etiolation and chlorosis in Spanish bluebells.

Sexual reproduction and variation

Sexual reproduction involves two parents, each contributing one gamete. There are a huge number of combinations of male and female gametes. A female gamete can be fertilised by any one of many male gametes. This is called **random fertilisation**.

Gametes are made by a type of cell division called **meiosis** (see page 43). During meiosis, either the maternal or the paternal chromosomes can be assorted to each side of the parent cell in order to make new daughter cells. This is called **independent assortment**.

Meiosis and crossing over

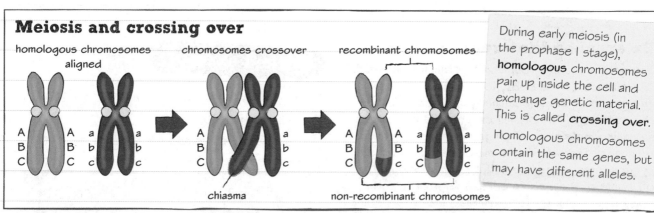

homologous chromosomes aligned

chromosomes crossover

recombinant chromosomes

chiasma

non-recombinant chromosomes

During early meiosis (in the prophase I stage), **homologous** chromosomes pair up inside the cell and exchange genetic material. This is called **crossing over**.

Homologous chromosomes contain the same genes, but may have different alleles.

Organisms that reproduce by asexual reproduction have limited variation. All daughter cells are clones, so variation can only come from **mutation**.

Worked example

How does sexual reproduction increase variation?

(3 marks)

Random fertilisation, independent assortment of chromosomes during meiosis and crossing over of genetic material during meiosis

Now try this

Skin colour gets darker when exposed to the sun. Describe and explain this type of variation.

(3 marks)

Inheritance

Monogenic inheritance

Monogenic inheritance is the inheritance of a single gene. For example, cystic fibrosis is a disease caused by a fault in the CFTR gene. There are two alleles for the CFTR gene. The **dominant** allele results in a normal phenotype. The **recessive** allele results in a disease phenotype. You can work out the pattern of inheritance using a genetic diagram:

The alleles from one parent are written on top of the square, and the alleles from the other parent are written on the side of the square. This is their genotype. Each allele combination is placed inside the square to work out the alleles of the offspring:

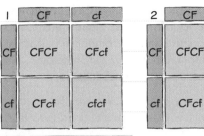

Key:
CF = dominant allele
cf = recessive allele

- two recessive alleles (homozygous recessive) = cystic fibrosis
- two dominant alleles (homozygous dominant) = no cystic fibrosis
- one of each allele (heterozygous) = carrier (no cystic fibrosis)

A carrier can pass on the recessive allele to their children.

Capital letters are used for dominant alleles; lower-case letters for recessive alleles.

Some monogenic phenotypes have multiple alleles. For example, the gene for blood group has three alleles, A, B, and O. A and B are dominant to O, and O is recessive. However, A and B are codominant. This means that if you inherit both A and B alleles, you will inherit both phenotypes.

Dihybrid inheritance

Dihybrid inheritance is the inheritance of two genes. For example, the genes for the shape (R or r) and colour (Y or y) of peas are inherited together.

When heterozygous parents are bred together, the phenotypes of the offspring always have the ratio 9:3:3:1.

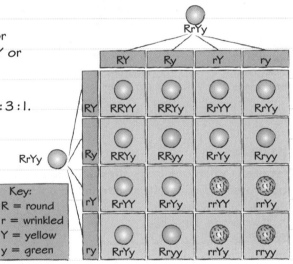

The allele for round peas (R) is dominant, and the allele for wrinkled peas (r) is recessive. The allele for yellow peas (Y) is dominant and the allele for green peas (y) is recessive.

Key:
R = round
r = wrinkled
Y = yellow
y = green

The possible genotypes of the mother are AA or AO.
The possible genotypes of the father are BB or BO.
If the mother is AO and the father is BO, four phenotypes, A, B, AB and O are possible.

Worked example

What are the possible phenotypes for offspring of a mother who is blood group A and a father who is blood group B? **(3 marks)**

blood type A (AO), blood type B (BO), blood type AB (AB), blood type O (OO)

Now try this

If 960 heterozygous pea plants were bred together, how many offspring would you expect to have green, smooth peas? **(2 marks)**

Linkage and epistasis

Autosomal linkage

Some genes are always inherited together because the loci for these genes are found together on the same chromosome. If they are found on a somatic (body) chromosome, then we describe this as **autosomal linkage**.

A condition called nail patella syndrome is caused by the dominant alleles for blood group and a protein that affects the nails and the patella, always being inherited together. If two people that are heterozygous for both alleles have children, the ratio of the phenotypes is not the expected 9:3:3:1. This indicates that the two genes are linked.

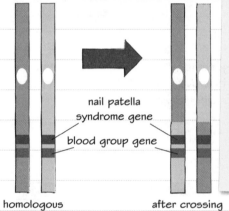

nail patella syndrome gene

blood group gene

homologous chromosomes

after crossing over

The loci for the nail patella gene and the blood group gene are close together on chromosome 9. After crossing over, the same alleles of these genes are still together.

Sex linkage

When genes on a sex chromosome (X or Y) are inherited, it is called **sex linkage**. One example of a sex-linked phenotype is haemophilia.

Haemophilia is the inability to clot blood effectively. It is caused by a single recessive allele of the haemophilia gene, found on the X chromosome.

XY
father unaffected by haemophilia

×

^HXX
mother carrier of haemophilia

Possible outcomes for each pregnancy

XY
25%
son without haemophilia

XX
daughter not a carrier

^HXY
25%
son with haemophilia

^HXX
daughter carrier of haemophilia

The mother is not affected by haemophilia but is a **carrier**.

Epistasis

Epistasis is when one gene affects the expression of another gene. This is often seen in the colours of flowers.

In this example, the colour of the petals is determined by two dominant alleles – C and P. If one of these alleles is recessive, the petals are white. If both of these alleles are dominant, the petals are purple. The offspring give a ratio of 9:7.

White variety #1 (CCpp)   ×   White variety #2 (ccPP)

F₁ generation

All purple (CcPp)

	CP	Cp	cP	cp
CP	CCPP Purple	CCPp Purple	CcPP Purple	CcPp Purple
Cp	CCPp Purple	CCpp White	CcPp Purple	Ccpp White
cP	CcPP Purple	CcPp Purple	ccPP White	ccPp White
cp	CcPp Purple	Ccpp White	ccPp White	ccpp White

Now try this

1. A man with haemophilia and a woman who does not have haemophilia, and is not a carrier, want to start a family. Draw a diagram to show the possible genotypes of the children and describe their phenotypes. **(3 marks)**

2. Describe and explain what would happen in a cross between a white flower (ccpp) and a purple flower (CcPp). **(4 marks)**

🖳 Maths skills　Using the chi-squared test

Chi-squared test example

The chi-squared test compares **observed** data (data we have collected) to the **expected** data (according to a theory). This is done in order to work out if the observed data are significantly different from those expected by the theory, or if the observed data are different only as the result of **chance**.

For example, two tall bean plants, that are heterozygous, are crossbred to produce 496 offspring (look back at page 139 for a reminder about monogenic and dihybrid inheritance). The allele for tall bean plants is dominant to the allele for short bean plants. Of these offspring, 387 are tall and 109 are short. Are these results significantly different from expected?

Our **null hypothesis** must be: There is no significant difference between the observed and expected values.

The formula for the chi-squared test is:

$$\chi^2 = \sum \frac{(\text{Observed} - \text{Expected})^2}{\text{Expected}}$$

Here are the data as a table:

	Observed (O)	Expected (E)	(O − E)	(O − E)^2	(O − E)^2/E
Number of tall plants	387	372	15	225	0.604839
Number of short plants	109	124	−15	225	1.814516
				$\Sigma =$	2.419355

We would expect the phenotypes of the bean plants to have the ratio 3 : 1, as 75% of the offspring would have at least one dominant allele. In this case, the ratio is 372 : 124. Putting these figures into the χ^2 formula, gives us a chi-squared value of 2.42.

To work out if this figure is significantly different, we must compare it to figures in a table of χ^2 critical values:

degrees of freedom ⟶

df	$p = 0.1$	$p = 0.05$	$p = 0.01$
1	2.706	3.841	6.635
2	4.605	5.991	9.210
3	6.251	7.815	11.345

We must compare our χ^2 value to the critical value in the table at $p = 0.05$. A probability value of 0.05 identifies the level that could occur by chance just 5 times in 100 (5%). We need to know that the probability of our results being the result of chance is greater than 5%.

We have one **degree of freedom**, because we have two phenotypes (n): degrees of freedom = n − 1.

If our χ^2 value is greater than 3.84, then our data are **statistically different** and we must **reject** our null hypothesis. Our χ^2 value is in fact less than 3.84, which means that we must **accept** our null hypothesis; that is, our results show no significant difference and this ratio of phenotypes is due to chance.

Worked example

What would it mean if the χ^2 value were 6.82?　　**(3 marks)**

There is a less than 1% probability that these results happened by chance.

Results are significantly different.

Reject null hypothesis.

Now try this

Using the same bean plant example from above, show what it would mean if the two heterozygous tall bean plants produced 392 tall offspring and 104 short offspring.

(4 marks)

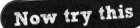

1% probability means that you would expect these results one time in 100 experiments.

$p = 0.05$ is the accepted level of significance in biology, but a χ^2 value that is $p = 0.01$ is better.

The evolution of a species

A species is a group of similar organisms that can interbreed to produce fertile offspring. Over time, species evolve by the process of natural selection. There are many factors that can affect the evolution of a species.

Selection pressure

When the environment is stable, the extremes of a phenotype are selected against:

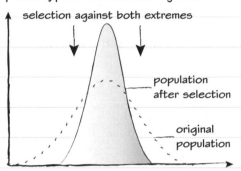

selection against both extremes

population after selection

original population

> This is called **stabilising selection**.

When the environment changes, one extreme is selected for:

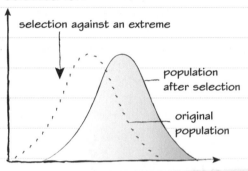

selection against an extreme

population after selection

original population

> This is called **directional selection**.

Sometimes, the phenotypes of a population change due to chance. For example, a natural disaster could wipe out a portion of the population, or a number of individuals could migrate elsewhere. This is called **genetic drift**.

The founder effect

When a population is suddenly reduced, for example due to a natural disaster, this is known as a **genetic bottleneck**.

The population will have a smaller number of individuals and less genetic variation.

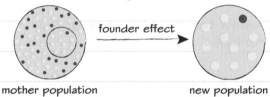

mother population   founder effect   new population

Isolating mechanisms and speciation

When two populations of a species are separated, for example on different islands, they will not be able to interbreed. This is **geographical isolation**. Each species will gradually change, due to mutations and natural selection, and eventually form a new species. This is called **speciation**.

- When speciation happens in different geographical areas it is called **allopatric** speciation.

- When speciation happens in the same geographical area it is called **sympatric** speciation.

Some individuals in a species may not be able to interbreed with other members of the same species. This is **reproductive isolation**.

> For example, Great Danes and Chihuahuas are both dogs, but would find it difficult to interbreed. Eventually these breeds of dog could become separate species.

Worked example

Explain why rabbits with short fur would be more common in a hot climate.   **(3 marks)**

Directional selection,

selects against rabbits with longer fur.

Rabbits with short fur more likely to survive.

Now try this

1. Explain how finches on the Galapagos Islands became separate species.   **(4 marks)**

2. Discuss the genetic variation of northern elephant seals. They were hunted almost to extinction in the 19th century, but have recovered their numbers from 20 individuals to 30 000.   **(3 marks)**

The Hardy–Weinberg principle

Maths skills The **Hardy–Weinberg** principle is an equation that allows you to work out the allele frequency of a population.

Assumptions of the principle

- the population size is large
- there is no migration, mutation or selection
- mating is random
- generations are non-overlapping
- organisms are diploid.

The equation is:

$p + q = 1$

p = dominant allele frequency

q = recessive allele frequency

It can also be written this way:

$p^2 + 2pq + q^2 = 1$

p^2 = frequency of individuals that are homozygous dominant

$2pq$ = frequency of individuals that are heterozygous

q^2 = frequency of individuals that are homozygous recessive.

Using the Hardy–Weinberg equation

The allele frequency of the population of flowers can be worked out using:

$p + q = 1$ **and** $p^2 + 2pq + q^2 = 1$

We know that 1 in 10 plants have a white flower and so have the genotype rr:
$q^2 = 0.1$.
Therefore q is the square root of
$q^2 : q = \sqrt{0.1}^2 = 0.316$.

We can use this information to work out the frequency of the genotypes of the red flowers. The red flowers can either be RR or Rr.

Using the first equation, $p = 1 - 0.316$
$= 0.684$

Using the second equation, p^2 is the square of $p : p^2 = (0.684)^2 = 0.468$

Again, using the second equation we know that
$2pq = 1 - q^2 - p^2 = 1 - 0.1 - 0.468$
$= 0.432$

We now know that 46.8% of the population are homozygous dominant (RR) and 43.2% of the population are heterozygous (Rr).

Half the plants leave offspring

In generation 1, a population of 1000 flowers has 1 white flower for every 9 red flowers. In generation 2, there are 3 white flowers for every 7 red flowers.

The recessive phenotype will always be homozygous recessive in this type of question.

Remember you can use both equations to work out the answer.

0.3 is the same as $\frac{3}{10}$; 0.548 can also be written as 54.8%.

Comparing the genotypes between generations 1 and 2 shows that genetic drift has occurred.

Worked example

If there are 1000 flowers in generation 2, how many of the red flowers in generation 2 have the genotype Rr?

(5 marks)

$\frac{3}{10}$ are white flowers, so $q^2 = 0.3$

$q = \sqrt{0.3} = 0.548$

$p = 1 - 0.548 = 0.452$

$p^2 = (0.452)^2 = 0.204$

$2pq = 1 - 0.3 - 0.204 = 0.496$, so 496 flowers have the genotype Rr.

Now try this

In the UK, one in every 2500 babies is born with cystic fibrosis. What percentage of the population are carriers? **(5 marks)**

Artificial selection

Selective breeding in plants and animals

Individuals with the desired characteristics are bred together and offspring with the desired characteristics are identified. Those offspring with the desired characteristics are allowed to breed, whereas offspring that do not have the desired characteristics are not. Over many generations, this can lead to very different looking breeds.

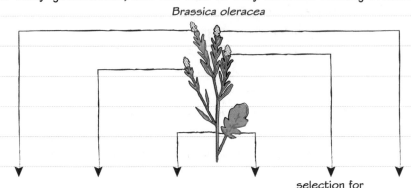

Brassica oleracea

For example, cabbage, broccoli, kale and cauliflower have all been selectively bred from wild cabbage, *Brassica oleracea*. This original plant is known as the **wild type**.

selection for terminal buds	selection for lateral buds	selection for stem	selection for leaved	selection for stems and flowers	selection for flower clusters
cabbage	brussels sprouts	kohlrabi	kale	broccoli	cauliflower

It is important to maintain as many wild-type plants as possible, because these are a source of genetic material. Recent projects include the Millennium Seed Bank Partnership at Kew, which has a seed bank of wild-type plants from around the world. These seeds have been collected to ensure that the plants do not become extinct due to climate change. They can be used in the future to make new crop varieties.

Ethical considerations of selective breeding

Selective breeding reduces the genetic variation in a population. It also increases the risk of a disease allele being inherited.

Many domesticated animals have health problems as a direct result of their breeding. For example, modern dog breeds have been selectively bred from the wolf. However, many of these breeds have breathing difficulties due to their short faces. Other breeds, such as the English bulldog, have a shortened life expectancy.

Worked example

Why are seed banks so important? **(3 marks)**

maintain biodiversity

a source of genetic variation

seeds can be used to make future plant varieties.

Now try this

A dog breeder wishes to breed only yellow Labrador puppies (homozygous recessive, ee). The breeder has several black (EE) and yellow (ee) male dogs and one chocolate (Ee) female dog. How could breeding yellow puppies be achieved? **(3 marks)**

Exam skills

This question revises patterns of inheritance. Look at pages 139, 140 and 143 to revise this.

Worked example

Petal colour is inherited by monogenic inheritance. Red petal colour (R) is dominant to white petal colour (r).

> The white petal colour is recessive to the red petal colour.

(a) What is meant by monogenic inheritance? **(1 mark)**

The inheritance of one gene.

> Heterozygous means that the organism has two different alleles.

(b) Two red heterozygous plants are bred together. Draw a genetic diagram to show the potential genotypes of the offspring.

(2 marks)

	R	r
R	RR	Rr
r	Rr	rr

> To draw this type of genetic diagram, put the genotype of one parent on the top of the table, and the genotype of the other down the side. Then place the alleles into the relevant squares, one from each parent.

(c) What is the ratio of the phenotypes red and white? **(1 mark)**

3:1

> From the genotypes in the genetic diagram, you can work out that one offspring in four, will have the genotype RR, two will have Rr and one will have rr. Since R is dominant to r, this means that three will be red, and one will be white.

(d) In a population, 14% of the plants have white flowers. Use the Hardy–Weinberg equations to work out what percentage of the population are homozygous dominant. **(4 marks)**

$$p + q = 1$$
$$p^2 + 2pq + q^2 = 1$$

$q^2 = 0.14$

Therefore, $q = \sqrt{0.14} = 0.374$

$p = 1 - 0.374 = 0.626$

$p^2 = 0.626^2 = 0.392$

39.2% of the population are homozygous dominant, RR.

> **Maths skills** q^2 is the frequency of individuals that are homozygous recessive, rr. q is the frequency of the recessive allele in the population, and p is the frequency of the dominant allele in the population. p^2 is the frequency of individuals that are homozygous dominant.

(e) Another plant controls petal colour by epistasis. Gene C controls the expression of the petal pigment. Gene P controls the colour, purple or pink. C means there is pigment and is dominant to c, no pigment. Purple colour, P, is dominant to pink, p.
What are the phenotypes of the petals on these plants?
(i) CcPp
(ii) Ccpp
(iii) ccPp **(3 marks)**

(i) purple petals
(ii) pink petals
(iii) white petals

> Epistasis means that the expression of one gene is controlled by another gene.
> For (i) and (ii), the C allele means that the pigment will be present. For (i) the P allele means that the colour will be purple. For (ii) the p means that the colour will be pink.
> For (iii) the c allele means that there will be no pigment. Whether the P or p allele are present, the petals will still be white.

DNA sequencing

DNA sequencing is the process of determining the order of the nucleotides in the DNA. Gene sequencing is DNA sequencing of the whole genome of an organism.

The principles of DNA sequencing

Primers are added to the template DNA to be sequenced. The primer binds to the start of the gene.

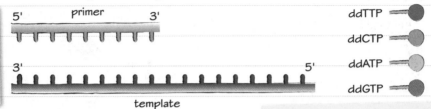

DNA polymerase enzyme, nucleotides and dideoxynucleotides (ddNTPs) are added to the template and primer. ddNTPs are labelled, each with a different fluorescent dye.

DNA polymerase makes many copies of the template. Whenever a ddNTP is incorporated, the sequence terminates. This results in many template copies, each with a different number of nucleotides.

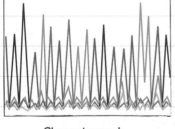

Chromatograph

New DNA sequencing techniques

Technology is advancing so that larger volumes of DNA can be sequenced faster and more cheaply. This is called **high-throughput sequencing**.

DNA sequencing is now automatic and carried out by computers. Many of the new techniques use DNA templates attached to microbeads or a solid surface. Many copies of the DNA template are made by the polymerase chain reaction (covered in more detail on page 147) in emulsion. These templates can be sequenced in parallel.

Cameras that can detect fluorescence or light emitted from the templates detect the sequence of nucleotides.

Whole **genomes** can now be sequenced in this way. A process that used to take years can now take hours.

A laser passes over each fragment as it passes the end of the gel. This excites the fluorescent dye and a detector detects the colour. The colours are recorded by a computer and used to produce a chromatograph.

Worked example

Explain how templates of different length are made by DNA sequencing. **(3 marks)**

DNA polymerase makes copies of the DNA template.

Nucleotides are incorporated into sequence.

Dideoxynucleotides / ddNTPs terminate sequences.

DNA polymerase is an enzyme that incorporates new DNA nucleotides into a copy of a DNA template.

Nucleotides are added in a **complementary** way – A to T, C to G.

ddNTPs prevent the next nucleotide from being added, so each sequence will end in a ddNTP.

Now try this

Scientists wish to examine which genes lead to coronary heart disease in the UK population. Which techniques and methods should they use? **(3 marks)**

Polymerase chain reaction

The polymerase chain reaction (PCR) is a method of making many copies of a DNA template.

PCR method

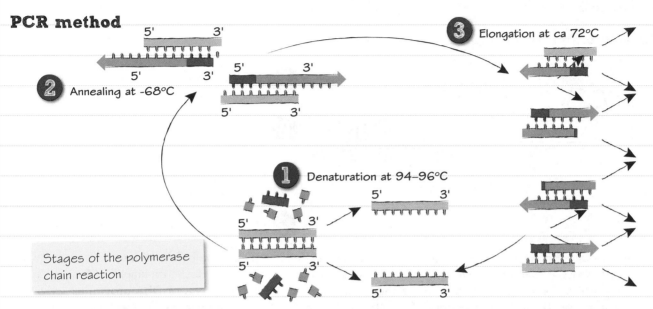

Stages of the polymerase chain reaction

The original DNA template is added to a mixture of nucleotides, primer and a type of DNA polymerase called **Taq polymerase**. A **primer** is a short section of DNA complementary to the beginning of the template. Taq polymerase can tolerate high temperatures and is from a bacteria called *Thermus aquaticus*. The mixture is added to a machine called a **thermocycler**.

- The thermocycler heats the mixture to 94−96 °C to separate the two DNA strands (**denaturation**).
- The mixture is then cooled to 68 °C to allow primers to bind to the template (**annealing**).
- The mixture is then heated to 72 °C to allow Taq polymerase to add nucleotides to the new strand of DNA (**elongation**).

You now have twice the amount of DNA strands that you had at the beginning. This process can be repeated until the required amount of DNA strands has been reached.

Uses of DNA profiling

PCR is often used in DNA profiling (or genetic fingerprinting) as only a small amount of DNA is usually acquired.

DNA profiling is used:

- to identify paternal or maternal relationships
- for personal identification
- in forensics
- to detect genetically inherited diseases.

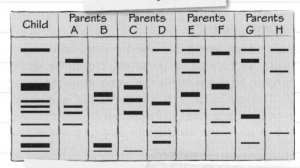

Genetic fingerprinting

Each individual will produce bands at different positions. Children share common bands with both parents.

Why must Taq polymerase be used in PCR?

(3 marks)

Human DNA polymerase would denature at high temperatures.

PCR uses high temperatures but Taq polymerase does not denature at 96 °C.

1 Use the genetic fingerprints above to decide which pair of parents are the parents of the child. **(1 mark)**

2 How many cycles of PCR does it take to make 262 144 DNA strands from one original template? **(2 marks)**

Gel electrophoresis

Gel electrophoresis is the process of separating DNA or proteins on a gel according to size and charge, using electricity.

How to carry out gel electrophoresis

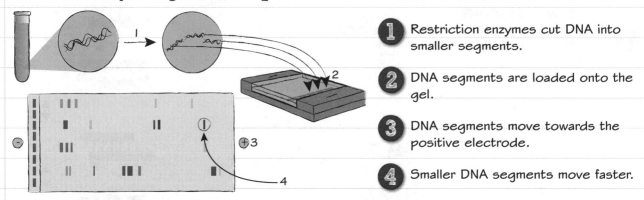

1 Restriction enzymes cut DNA into smaller segments.

2 DNA segments are loaded onto the gel.

3 DNA segments move towards the positive electrode.

4 Smaller DNA segments move faster.

A DNA 'ladder' containing bands of known size is often added to the first well. This allows the scientists to work out the size of each DNA band. If there is only a small amount of DNA, it can be amplified first using PCR (page 147). The DNA bands can be identified using a dye, UV fluorescence or DNA probes specific to a particular gene. The DNA on the gel can also be transferred onto a membrane using a technique called Southern blotting. The DNA bands can be identified on the membrane using radioactive DNA probes.

Applications

- Each person's DNA contains sections of repeated DNA sequences called variable number tandem repeats (VNTR). These are different in length and number in each person.

- If the DNA is cut in each VNTR by a restriction enzyme, it will result in a number of DNA strands of different lengths.

- When these strands are separated by gel electrophoresis, each person will have a unique pattern of bands. This is the principle behind genetic fingerprinting.

Gel electrophoresis can also be used to identify a gene of interest, or to analyse proteins. Proteins can be separated according to mass using the same principle as separating DNA. The gel is run in the presence of sodium dodecyl sulfate (SDS), a charged detergent. Without the SDS, proteins are separated according to surface charge.

Practical skills | **Using gel electrophoresis**

You may be asked to carry out gel electrophoresis in a lesson, using a small agarose gel in an electrophoresis tank with buffer. A power pack will run electricity through the gel. Take care to get your entire sample into the well.

Worked example

Why is each person's genetic fingerprint different?   **(4 marks)**

Different numbers and lengths of VNTRs mean that restriction enzymes will cut different lengths of DNA. Gel electrophoresis will separate these lengths as bands on the gel.

Children inherit half of their VNTRs from their mother and half from their father. This means that you are able to compare DNA from parents and offspring.

Restriction enzymes are also called **restriction endonucleases**.

Each restriction enzyme cuts the DNA is a specific place.

Now try this

Describe how you would identify a section of DNA of interest using gel electrophoresis.   **(4 marks)**

Genetic engineering

Genetic engineering is the use of technology to change the genetic material of an organism.

Principles

When the genetic material of an organism is altered, the organism is described as being **genetically modified**. When a gene from an organism of one species is placed into an organism of a different species, the organism can be described as **transgenic**.

Genes can be isolated from the DNA of an organism by extracting the DNA and cutting out the gene using restriction enzymes. Genes can also be isolated by extracting the messenger RNA (mRNA) of the gene of interest from cells, and using an enzyme called **reverse transcriptase** to convert the mRNA into DNA. The gene can be placed into the cells of the other organism using a **vector**, such as a plasmid.

Techniques

Restriction enzymes only cut the DNA at specific sites. Using the right restriction enzymes, you can cut the DNA either side of your gene of interest, for example insulin.

The **plasmid** vector must also be cut using the same restriction enzymes, so that **sticky ends** are made. These are overlapping sections of DNA that can be joined together using an enzyme called **DNA ligase**. The resulting plasmid is said to contain **recombinant DNA** (DNA from two different species).

Plasmids can be taken up by cells using **electroporation**. A small electrical charge is placed over the cells, which increases the permeability of the cell surface membrane. The plasmids can cross the membrane and enter the cells.

Making human insulin in E. coli

1 The human insulin gene is cut out of the DNA using restriction enzymes.

gene for human insulin
sticky

2 A plasmid is removed from an *E. coli* bacterial cell and cut open using the same restriction enzymes.

E.coli cell
bacterial chromosome
restriction enzyme
plasmid

3 The restriction enzymes create sticky ends either side of the insulin gene and in the plasmid.

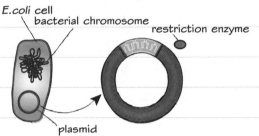

DNA recombination

sticky ends

4 The human insulin gene is placed into the *E. coli* plasmid. The sticky ends are joined together using DNA ligase.

recombinant plasmid
gene for human insulin

5 *E. coli* cells take up the recombinant plasmid after electroporation. The *E. coli* cells can now make human insulin.

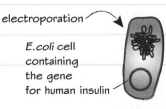

electroporation

E.coli cell containing the gene for human insulin

Worked example

Discuss the roles of enzymes in genetic engineering. **(3 marks)**

DNA is cut using restriction enzymes. Sticky ends are joined using DNA ligase. Reverse transcriptase is used to convert mRNA to DNA.

Now try this

Describe how genetic engineering could be used to place the gene for human blood clotting factor into goats. **(5 marks)**

The ethics of genetic manipulation

There are ethical reasons to genetically modify organisms, as well as ethical reasons not to. You should be aware of both sides of the argument.

Animals and microorganisms

We use genetically manipulated animals and microorganisms on an everyday basis. For example, genetically modified goats produce human blood clotting factor in their milk. When animals are used to produce pharmaceutical products in this way, it is known as 'pharming'. However, there are many ethical issues to consider.

Genetically manipulated organism	Ethical issue
Genetically modified *E. coli* producing human insulin	Diabetics receive human insulin, reducing human pain and suffering
Genetically modified pigs that express higher levels of growth hormone	Pigs grow too fast, causing heart discomfort when exercising
Genetically modified goats that produce human lysozyme	Protects humans who drink the goat milk from diarrhoea
Planning to clone northern white rhino	To prevent extinction of northern white rhino
Cloned cattle	Increased susceptibility to disease in cattle
Children born with SCID (severe combined immunodeficiency) treated with gene therapy	Children cured of SCID, removing pain and suffering
Genetically modified 'oncomouse' patented	Patented animals as human property

Plants

We use plant cloning and genetically modified plants to increase productivity in farming. In recent years, the safety of genetically modified crops has been called into question, and ethical issues surrounding the production of patented seeds have been raised.

Genetically manipulated organism	Ethical issue
Genetically modified rice contains beta carotene genes	People who eat genetically modified rice less likely to suffer from vitamin A deficiency
Genetically modified tomatoes that take longer to soften after ripening	Concerns over safety of eating genetically modifed food
Genetically modified maize that is resistant to drought	Greater yield of maize in countries with drought, feeding more people and averting famine
Genetically modified soya contains gene that makes it resistant to insects	If genetically modified soya crossbreed with wild plants, resistance gene could be spread to weeds
Genetically modified tobacco plants that produce therapeutic antibodies	Therapeutic antibodies used to treat rabies and save human lives
Patented genetically modified seeds	Farmers cannot use seed from crops and must buy more seed from manufacturer

Worked example

Discuss any ethical concerns of using genetically modified rice. **(4 marks)**

Eating genetically modified rice will lead to fewer health problems caused by vitamin A deficiency.

There are concerns over safety of eating GM rice.

Genetically modified rice may crossbreed with wild-type rice and spread genes.

Farmers must buy seed each year instead of using seed from crop.

Genetically modified rice has been developed for growth in areas of the world where vitamin A deficiency is a major health problem.

Seeds can cross-pollinate across some species. There is always a chance that this can happen if you plant seeds outdoors.

This will affect poor farmers in particular.

Now try this

1 Should we genetically engineer animals? Give arguments for and against.   **(4 marks)**

2 What ethical considerations should be taken into account when cloning plants?   **(4 marks)**

Gene therapy

Gene therapy is the replacement of a faulty gene with a functional gene.

Principles of gene therapy

Many diseases are caused by the inheritance of two recessive 'faulty' alleles. An example is cystic fibrosis.

If a dominant functional gene can be introduced to the cells, then the disease can be reversed.

There are two types of gene therapy:

- germ line cell gene therapy – the functional allele is added to the gametes. All somatic (body) cells will contain the functional allele, and it could be passed onto offspring. This is currently banned in the UK and other countries.

- somatic cell gene therapy – the functional allele is added to the somatic cells. The functional allele could not be passed onto offspring.

Uses of gene therapy in medicine

Gene therapy could be used in the future to treat any recessive single-gene disease. It could also be used to alter or kill aberrant cells, for example cancer cells, or induce cells to produce therapeutic proteins, such as producing fetal haemoglobin in patients with sickle cell anaemia.

Gene therapy has been used successfully to treat patients with severe combined immunodeficiency (SCID). These patients have no functional immune system, caused by a fault in the ADA gene.

Problems with gene therapy include:

- identifying an appropriate target
- getting the functional allele into the right cells
- delivering the functional allele in a vector that does not trigger an immune response

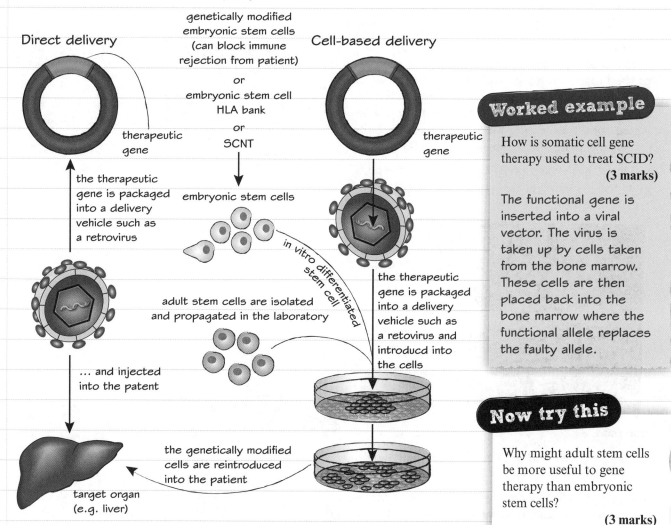

Direct delivery

genetically modified embryonic stem cells (can block immune rejection from patient)

or

embryonic stem cell HLA bank

or

SCNT

Cell-based delivery

therapeutic gene

therapeutic gene

the therapeutic gene is packaged into a delivery vehicle such as a retrovirus

embryonic stem cells

in vitro differentiated stem cell

adult stem cells are isolated and propagated in the laboratory

the therapeutic gene is packaged into a delivery vehicle such as a retovirus and introducd into the cells

... and injected into the patent

the genetically modified cells are reintroduced into the patient

target organ (e.g. liver)

Worked example

How is somatic cell gene therapy used to treat SCID? **(3 marks)**

The functional gene is inserted into a viral vector. The virus is taken up by cells taken from the bone marrow. These cells are then placed back into the bone marrow where the functional allele replaces the faulty allele.

Now try this

Why might adult stem cells be more useful to gene therapy than embryonic stem cells?

(3 marks)

Exam skills

This question uses knowledge and skills that you have already revised. Look at pages 147 and 148 to revise gene techniques.

Worked example

(a) Forensic scientists have recovered a small sample of DNA from a crime scene.

   (i) What technique could be used to increase the amount of DNA in the sample? **(1 mark)**

polymerase chain reaction (PCR).

   (ii) Describe how to carry out this technique. **(4 marks)**

Mix the DNA template with primers, DNA nucleotides and DNA polymerase. Heat the DNA template to 95 °C to separate the two strands. Cool the mixture to 68 °C to allow the primers to anneal to the two strands. Heat the mixture to 72 °C to allow the new DNA strands to elongate.

(b) The DNA sample is compared to some suspects' DNA using gel electrophoresis. Explain how this works. **(4 marks)**

The DNA sample and the suspects' DNA are digested using restriction enzymes. The samples are then loaded onto an agarose gel and covered with buffer in a tank. Electricity is passed through the tank, separating the DNA fragments according to size. Small fragments will move quickly through the gel and large fragments will move more slowly.

(c) This is the pattern of the DNA fragments, or bands, on the gel. Which of the suspects left their DNA at the crime scene? **(1 mark)**

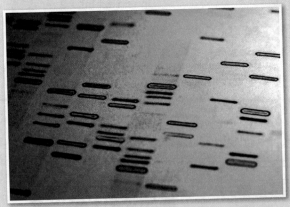

Suspect 2

(d) Why does each person have a different pattern of DNA bands? **(2 marks)**

Each person's DNA contains sections of repeated DNA sequences called variable number tandem repeats (VNTR) that are different in length and number.

Practical skills PCR is carried out inside a machine called a thermocycler, which can heat and cool the mixture.

Heat is required to break the hydrogen bonds between the complementary base pairs.

A primer is a small section of DNA that is complementary to one end of the DNA template.

The polymerase used in PCR is from *Thermus aquaticus*, a type of bacteria found in hot springs that can withstand high temperatures.

Practical skills The restriction enzymes digest the DNA wherever there is a VNTR sequence. This breaks the DNA sample into smaller fragments.

DNA has a slight negative charge. The DNA fragments move towards the positive anode when electricity is passed through the gel.

The DNA bands for suspect 2 match the bands from the crime scene DNA.

The restriction enzymes will therefore digest the DNA in different places to produce different size fragments. After gel electrophoresis this produces a unique pattern of bands.

Natural clones

Clones are genetically identical organisms. Many plants are natural clones.

Natural clones in plants

Many plants reproduce **asexually** using a form of **vegetative propagation**. There is only one parent and all of the offspring are genetically identical to each other and the parent.

For example, strawberry plants reproduce by sending out horizontal stems called **stolons**. When a stolon touches the ground, new roots grow and eventually a new plant grows.

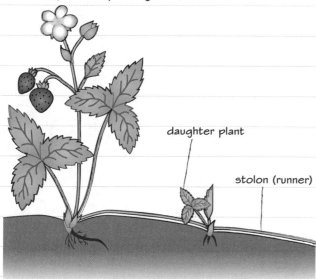

daughter plant

stolon (runner)

Potatoes are a type of **tuber**, an enlarged stem for food storage. Each potato produces many small growths on their surface called sprouts that can grow into a new potato plant.

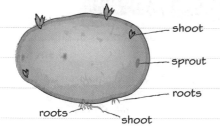

shoot

sprout

roots

roots

shoot

Daffodils and other flowers come from bulbs. Each year, the outer layer of the bulb grows into a flower.

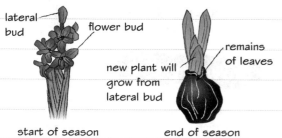

lateral bud

flower bud

remains of leaves

new plant will grow from lateral bud

start of season

end of season

Natural clones in animals

If an embryo splits during early development, two genetically identical embryos will form and grow into twins. These are natural animal **clones**.

How to take plant cuttings

The natural ability of plants to form clones is used in horticulture. New plants can be grown by taking a **cutting** from the parent plant and placing it into soil. Sometimes plant **hormones** are added to the base of the cutting to encourage growth.

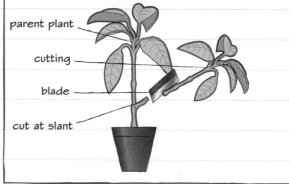

parent plant

cutting

blade

cut at slant

Worked example

What is meant by vegetative propagation? **(3 marks)**

asexual reproduction in plants

new plants grow from tubers / bulbs / stolons

Offspring are genetically identical to parent.

Asexual means 'without sex'.

New plants can also grow from cuttings and grafting.

Genetic variation can be introduced into the daughter plants by mutation.

Now try this

1   What are the advantages of cloning plants?

(4 marks)

2   Explain why fraternal twins are not clones.

(2 marks)

Artificial clones

Artificial clones are clones that have been made by *in vivo* or *in vitro* cloning techniques.

Artificial cloning in plants

Micropropagation can be used to produce many plants from one parent plant:

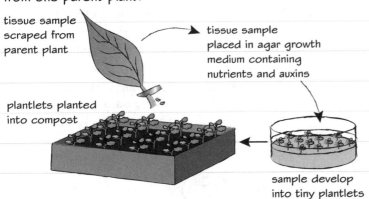

tissue sample scraped from parent plant

tissue sample placed in agar growth medium containing nutrients and auxins

plantlets planted into compost

sample develop into tiny plantlets

Advantages and disadvantages

	Advantages	Disadvantages
Artificial plant cloning	All of the plants have the desired characteristics All of the plants can be grown in similar conditions	No genetic variation All of the plants will be susceptible to the same diseases
Artificial animal cloning	Animal has the desired characteristics Animals that can produce pharmaceuticals in their milk could be quickly reproduced	No genetic variation Health problems in cloned animals Expensive procedure

Artificial cloning in animals

Animals can be cloned using one of two techniques: artificial embryo twinning and somatic cell nuclear transfer (SCNT).

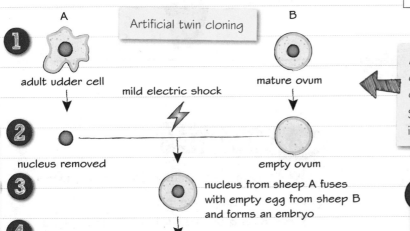

Artificial twin cloning

A

B

① adult udder cell   mature ovum

mild electric shock

② nucleus removed   empty ovum

③ nucleus from sheep A fuses with empty egg from sheep B and forms an embryo

④ cloned embryo implanted into sheep C

⑤ C
lamb born is a clone of sheep A

Artificial twin cloning is used when many copies of an animal with the desired characteristics is wanted.

SCNT is used when an animal genetically identical to an existing animal is wanted.

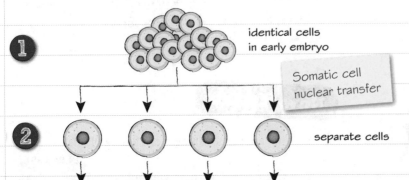

① identical cells in early embryo

Somatic cell nuclear transfer

② separate cells

③ cells develop into identical embryos

④ identical clones born

Worked example

Compare the two methods of animal cloning.   **(3 marks)**

Artificial embryo cloning uses a fertilised ovum and SCNT uses an unfertilised ovum to make an embryo.

Artificial embryo twinning makes many clones but SCNT makes one.

The ovum in SCNT contains a nucleus from a somatic cell.

Now try this

What are the disadvantages of artificial cloning?   **(5 marks)**

Microorganisms in biotechnology

Biotechnology is the use of organisms in technology to make a useful product.

Microorganisms and fermenters

Biotechnology often uses microorganisms such as bacteria or fungi. These can be grown in bulk inside fermenters.

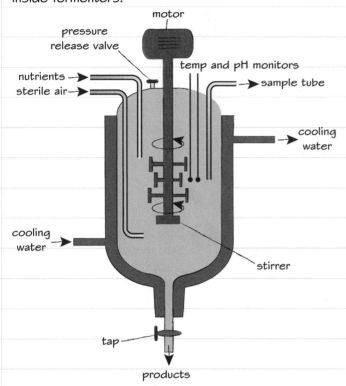

- To achieve the maximum yield, the microorganisms must be kept at the optimum temperature and pH. This is accomplished using the water-cooling jacket and temperature and pH probes.
- Microorganisms are grown in nutrient broth, which contains the sugars and amino acids needed for growth.
- Oxygen for aerobic respiration is added to the fermenter.
- The fermenter is mixed using stirrers.
- The contents of the fermenter are kept sterile to avoid contamination with unwanted microorganisms.

Uses of microorganisms

Microorganisms are used in:

- brewing (yeast)
- baking (yeast)
- cheese making (*Lactobacillus* bacteria)
- yoghurt production (*Lactobacillus* bacteria)
- mycoprotein production (*Fusarium* fungi)
- soya sauce fermentation (*Aspergillus* fungi)
- penicillin production (*Penicillium* fungi)
- insulin production (*E. coli* bacteria)
- bioremediation (the breakdown of waste substances).

Microorganisms in food production

Advantages	Disadvantages
fast growth rate can be grown on waste products easy to set up and maintain can quickly respond to demand.	possible contamination with a different microorganism unwanted microorganisms could affect the health of workers people might not like the idea / taste.

Now try this

1 Suggest what would happen to a fungus growing in the fermenter if the pH fell below the optimum level. **(3 marks)**
2 What might happen in a fermenter if the bacteria did not get enough oxygen? **(2 marks)**

Worked example

What are the advantages of using genetically modified E. coli to make human insulin? **(4 marks)**

E. coli have a fast growth rate.

All of the bacteria will produce human insulin.

Easy to set up and maintain bacteria.

No ethical concerns about using bacteria.

E. coli are commonly used in biotechnology because they reproduce asexually by binary fission every 20 minutes.

E. coli are easy to genetically engineer using plasmids (see page 149).

Some people may have ethical concerns about using insulin from animals.

Aseptic techniques

Asepsis means 'without microorganisms'. Aseptic techniques prevent the growth of unwanted microorganisms.

What are aseptic techniques?

- sterilising equipment
- sterilising solids, liquids and gases
- hand washing
- wearing protective clothing
- maintaining an air pressure difference between the air in the room and outside, so that air flows out of the room.

Batch and continuous fermentation

Batch fermentation is when microorganisms are set up in a fermenter and nothing is added or taken away (except for waste gases) until the fermentation is finished.

Penicillin production is batch fermentation. Penicillin is only produced at a later stage of the process as a **secondary metabolite**.

Continuous fermentation is when nutrients are steadily added to the fermenter and there is a steady harvest of the product.

Mycoprotein production is continuous. Mycoprotein is a **primary metabolite** and is produced throughout the process.

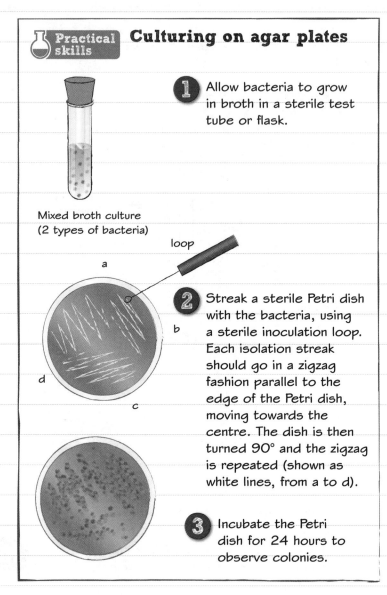

Practical skills | **Culturing on agar plates**

1 Allow bacteria to grow in broth in a sterile test tube or flask.

Mixed broth culture (2 types of bacteria)

loop

2 Streak a sterile Petri dish with the bacteria, using a sterile inoculation loop. Each isolation streak should go in a zigzag fashion parallel to the edge of the Petri dish, moving towards the centre. The dish is then turned 90° and the zigzag is repeated (shown as white lines, from a to d).

3 Incubate the Petri dish for 24 hours to observe colonies.

Worked example

Why are aseptic techniques important? **(3 marks)**

Unwanted microorganisms could compete with microorganisms in the fermenter and decrease the yield.

Protect the health of workers.

Unwanted microorganisms may produce toxic chemicals.

If unwanted microorganisms get into the fermenter, they will use up some of the nutrient broth and oxygen, which would reduce production of the product. They could also affect the taste of the product.

Unwanted microorganisms may be harmful to health.

Toxic chemicals would make the product unusable.

Now try this

1 Explain why aseptic conditions are more important in continuous fermentation. **(2 marks)**
2 Describe what precautions you should take when culturing bacteria on agar plates. **(5 marks)**

Growth curves of microorganisms

All microorganisms follow the same standard growth curve if grown in a closed culture.

Standard growth curve of microorganisms

1 **Lag phase** – Bacteria adjust to new conditions and synthesise new proteins. Slow growth.

2 **Log phase** (exponential growth phase) – The number of bacteria double in each unit of time. No restrictions to growth, so growth is rapid.

3 **Stationary phase** – Death rate of bacteria equals the division rate. Growth is limited by the amount of nutrients and oxygen available, and the build up of waste products.

4 **Death phase** (logarithmic decline phase) – Death rate of bacteria is greater than the division rate due to lack of nutrients or toxic build up of waste products.

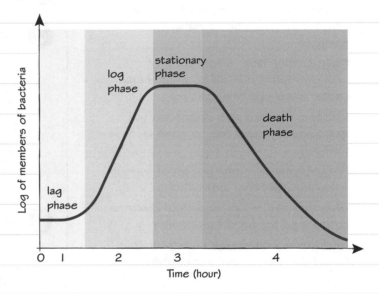

Factors affecting growth

- amount of nutrients
- amount of space
- oxygen availability
- water availability
- temperature
- pH
- toxic waste products.

Worked example

Why do the bacteria grow slowly during the lag phase?   **(3 marks)**

adjust to new conditions / food sources

Genes need to be switched on

New proteins must be synthesised.

For example, if *E. coli* find themselves in a lactose-only environment, they will need to switch on the genes in the lac operon (look back at page 135 as a reminder).

Beta-galactosidase and lactose permease have to be synthesised.

Practical skills **Serial dilutions**

Solutions of bacteria are diluted so that they can be plated onto agar plates for counting.

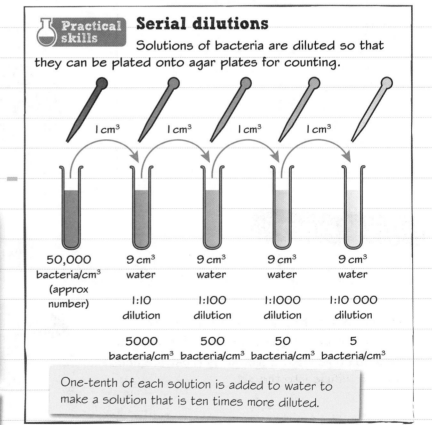

One-tenth of each solution is added to water to make a solution that is ten times more diluted.

Now try this

If a 1:10 000 dilution contains 28 bacteria per cm³, how many bacteria are there in 10 cm³ of the original solution?   **(3 marks)**

Immobilised enzymes

Enzymes used in industry are often immobilised to allow them to be reused.

Uses of immobilised enzymes

- Glucose isomerase — conversion of glucose to fructose to make food sweeter.
- Penicillin acyclase — formation of semi-synthetic penicillins, which some bacteria are not resistant to.
- Lactase — hydrolysis of lactose to glucose and galactose to make lactose-free milk.
- Aminoacyclase — production of L-amino acids for industrial applications.
- Glucoamylase — conversion of dextrins to glucose for food sweetening.
- Nitrilase — conversion of acrylonitrile to acrylamide for use in the plastics industry.

Reasons for immobilising enzymes

- To reuse the enzymes many times.
- The enzymes will not contaminate the product.
- Enzymes are more tolerant of temperature and pH changes.
- Enzyme activity can be controlled more accurately.

Methods of immobilisation

1 Adsorption – enzyme is bound to a carrier such as activated carbon, clay or glass beads.

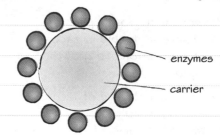
enzymes
carrier

2 Entrapment – enzyme is trapped inside a gel such as alginate beads.

enzymes
gel

3 Cross-linking – amino acids on the enzymes are linked together using glutaraldehyde.

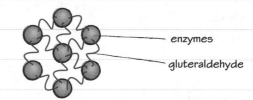
enzymes
gluteraldehyde

What are the advantages and disadvantages of immobilising enzymes inside alginate beads, compared to binding enzymes to a carrier? **(3 marks)**

Enzyme is not likely to contaminate product.

Enzyme will be more tolerant of temperature and pH.

Reaction time may be slower as substrate needs time to diffuse into the gel.

Adsorption and cross-linking both carry a risk of **leakage**, when the enzyme becomes detached.

The gel surrounds the enzymes and gives protection from increases and decreases in temperature and pH.

The reaction times in adsorption and cross-linking will be much quicker.

1 Describe how immobilised lactase could be used to remove the lactose from milk. **(4 marks)**
2 Hypothesise what could happen to cross-linked enzymes if the temperature in the reaction increased by 10 °C. **(3 marks)**

Exam skills

This question is about the use of microorganisms in biotechnology. Look at pages 155–158 to revise these topics.

Worked example

(a) Name two uses of microorganisms in biotechnology. **(2 marks)**

One food example, e.g. cheese / brewing

One medical example, e.g. penicillin / insulin

> Food examples also include baking, yoghurt production, mycoprotein production and soya sauce fermentation.
>
> Microorganisms can also be used in bioremediation, the break down of waste substances.

(b) What are the advantages of using microorganisms in biotechnology? **(4 marks)**

A fast growth rate, can be grown on waste products, cheap, and easy to set up and maintain fermenters.

> Fermenters can be set up anywhere and can quickly respond to demand.

(c) Fermenters can be used to culture large quantities of microorganisms. What is the difference between batch fermentation and continuous fermentation? **(2 marks)**

In batch fermentation, microorganisms are set up in a fermenter and nothing is added or taken away until fermentation is finished. In continuous fermentation, nutrients are steadily added to the fermenter and there is a steady harvest of the product.

> The advantage of batch fermentation is that it allows secondary metabolites to be produced, e.g. penicillin.
>
> The advantage of continuous fermentation is that fermentation does not need to stop.

(d) Fermentation must be carried out under aseptic conditions. What does asepsis mean? **(1 mark)**

without microorganisms

> **Practical skills** Fermenters are kept aseptic by sterilising all equipment. Everything that goes into the fermenter must also be sterile.

(e) Explain why asepsis is important. **(2 marks)**

Unwanted microorganisms could compete with microorganisms in the fermenter and decrease the yield. Unwanted microorganisms may produce toxic chemicals.

> This could lead to the product having an unpleasant taste.

(f) Describe the standard growth curve of microorganisms in a fermenter. **(4 marks)**

The growth curve starts with slow growth called the lag phase, where the bacteria adjust to new conditions and synthesise new proteins, followed by a rapid growth phase, called the log phase. Eventually, the death rate of bacteria equals the division rate. This is called the stationary phase. When the nutrients run out, or there is a toxic build up of waste products, the death rate of bacteria is greater than the division rate. This is called the death phase.

> An annotated graph would be an alternative way to answer this question. This is the basic shape of the graph.

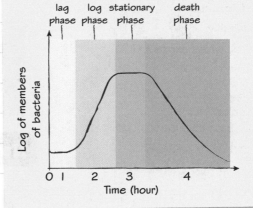

159

Ecosystems

An ecosystem is all the living and non-living components in a habitat and their interactions.

Biotic and abiotic factors

An ecosystem contains both biotic and abiotic factors.

Biotic factors are living components of the ecosystem, for example:

- **interspecies** competition
- **intraspecies** competition
- predation disease
- food supply.

Abiotic factors are non-living components of the ecosystem, for example:

- light intensity
- pH of the soil
- water availability
- air movement
- oxygen availability.

Ecosystems are **dynamic**; all of the components interact with each other and influence the size of the populations within the ecosystem.

Rock pool ecosystem

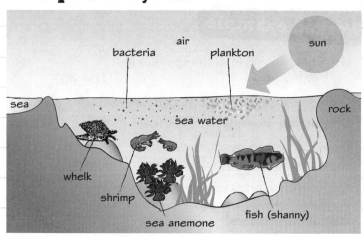

Playing field ecosystem

A playing field has few species as it is regularly mowed. However, there are several species of grass and some other small flowering plants, for example daisies.

Biotic:

- interspecies competition for sunlight
- intraspecies competition for minerals.

Abiotic:

- water availability
- mineral availability.

Large tree ecosystem

Biotic:

- predation
- disease.

Abiotic:

- pH of the soil
- water availability.

branches and leaves: bees, wasps, moths, squirrels, sparrows and hawks

trunk: insects and larvae

root and litter zone: bacteria, earthworms, wood lice and fungi

Worked example

Which abiotic factors affect plants living in a rock pool?   **(4 marks)**

light intensity

salinity of the water

water availability

temperature, when exposed

Only choose abiotic factors that will affect the rock pool plants specifically. Be careful not to choose any biotic factors.

Now try this

A new pesticide is sprayed close to a large tree and the number of insects in the tree decreases. Explain what would happen to the number of blue tits.   **(3 marks)**

Biomass transfer

Each stage of a food chain is called a **trophic level**. We can measure the biomass transfer as it flows through each trophic level of an ecosystem.

Biomass transfers through ecosystems

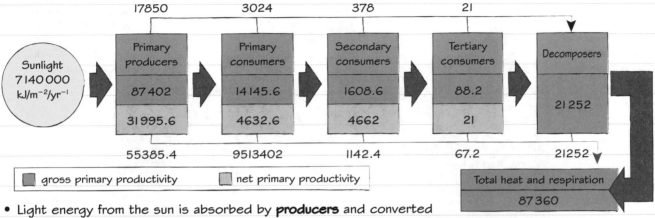

- Light energy from the sun is absorbed by **producers** and converted into chemical energy. This is the producer's **gross primary productivity** (GPP).
- Some of this energy is lost as thermal (heat) energy due to **aerobic respiration**.
- The remaining energy goes into the biomass of the producer. This is the **net primary productivity** (NPP).
- Some of the NPP is transferred to the **primary consumer** when it is consumed, or the **decomposers** if the plant dies.

As each consumer feeds, some of the biomass from the previous trophic level cannot be consumed or will be lost as egestion. Some energy will be lost due to respiration and excretion. For this reason, there is less biomass and less energy available for each successive trophic level. This is why there are not usually more than five or six trophic levels in an ecosystem.

Measuring the productivity

One method of measuring the **dry biomass** of each trophic level is to remove the water from one organism in the trophic level. One gram of the biomass is burned in a **calorimeter** to measure the energy content. The total energy content in the population is worked out using the formula:

$$\text{energy content} = \text{number of organisms} \times \text{mean dry mass of one organism} \times \text{energy content of 1 g of dry mass}$$

The GPP and the NPP can be worked out in this way. The units for the total energy content are kJ m^{-2} year^{-1}.

The efficiency of energy transfers can be worked out using the formula:

$$\text{efficiency} = \frac{\text{useful energy out}}{\text{energy in}} \times 100\%$$

Manipulating biomass transfer

In farming and horticulture, a greater **yield** can be achieved if there is more biomass to transfer. It is possible to:

- use artificial light to increase hours of light and increase the rate of photosynthesis
- increase the temperature, using greenhouses, to increase the rate of photosynthesis
- restrict the movement of animals so that less energy is lost as thermal energy due to aerobic respiration
- carefully control animal feed.

Some parts of the secondary consumer cannot be consumed, for example fur. Not all of the animal will be consumed due to inefficient eating.

Egestion is the removal of faeces. Excretion is the removal of urine and waste gases.

Worked example

How is energy lost from the secondary consumers? **(3 marks)**

thermal energy from aerobic respiration

chemical energy lost by egestion

chemical energy lost by excretion

Now try this

1 If the biomass of the producers is 87000 kg and the biomass of the tertiary consumers is 88 kg, what is the efficiency of the ecosystem? **(2 marks)**

2 Some farmers keep their livestock in enclosed pens. Explain why. **(2 marks)**

Nitrogen cycle

Nitrogen is cycled through an ecosystem using several different groups of microorganisms.

Atmospheric nitrogen can be fixed into ammonium by nitrogen-fixing bacteria in the soil or in the roots of **leguminous plants**, e.g. pea plants, clover.

Nitrates are taken up by the plants and used to make amino acids and proteins. Animals can take up nitrogen compounds in the form of proteins when they consume plants.

Plant and animal proteins (organic nitrogen compounds) are returned to the soil when the organisms die, or from animal waste, and are converted into ammonium once again.

nitrogen in atmosphere (N_2)

plants

assimilation

nitrogen-fixing bacteria in root nodules of legumes

decomposers (aerobic and anaerobic bacteria and fungi)

denitrifying bacteria

nitrates (NO_3^-)

Nitrates are converted into atmospheric nitrogen.

ammonification

nitrification

nitrifying bacteria

nitrogen-fixing soil bacteria

ammonium (NH_4^+)

nitrifying bacteria

nitrites (NO_2^-)

Some plants can use ammonium directly; others can only use it in the form of nitrates. Ammonium is converted to nitrites and then nitrates by nitrifying bacteria in the soil.

Microorganisms in the nitrogen cycle

- There are two types of **nitrogen-fixing bacteria**: *Azotobacter* fixes atmospheric nitrogen into ammonium in the soil. *Rhizobium* fixes atmospheric nitrogen into ammonium in the soil and the nodules of leguminous plants.
- **Decomposers** for example bacteria and fungi – decompose dead animal and plant matter into ammonium.

- There are two types of **nitrifying bacteria** – *Nitrosomonas* converts ammonium in the soil into nitrites. *Nitrobacter* converts nitrites in the soil into nitrates.
- **Denitrifying bacteria** – convert nitrates in the soil into atmospheric nitrogen.

How do plants obtain nitrates from the soil? **(3 marks)**

Nitrogen fixation of atmospheric nitrogen by nitrogen-fixing bacteria.

Ammonium is converted into nitrates by nitrifying bacteria.

Plants absorb nitrates from the soil.

1 Explain the significance of *Nitrosomonas* and *Nitrobacter* for the growth of plants. **(3 marks)**
2 Why is adding fertiliser to fields often necessary to obtain a high yield of crops?
(4 marks)

Carbon cycle

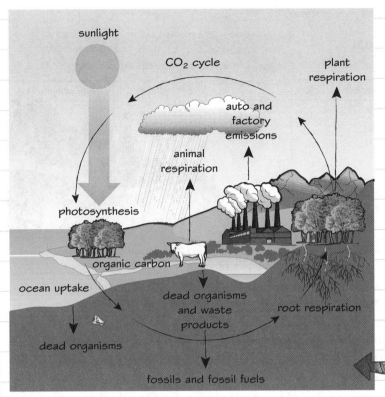

- Most carbon dioxide is removed from the atmosphere by photosynthesis.
- It can be stored under the ground as fossil fuel (coal, oil, gas) when organisms die.
- Carbon dioxide has been absorbed by the oceans and stored in sedimentary rocks when sea organisms die.
- Carbon dioxide is placed back into the atmosphere by aerobic respiration, decomposition, combustion and the eruption of volcanoes.

Carbon is cycled through an ecosystem by the actions of organisms.

All of these effects are due to an increased number of humans on the planet. We all need energy, and space to live and grow food.

Remember to state **aerobic** respiration, as anaerobic respiration does not give off carbon dioxide.

Processes within the carbon cycle

- **Photosynthesis:** Plants and photosynthetic bacteria fix carbon from carbon dioxide into glucose and other carbon-containing compounds.
- **Aerobic respiration:** carried out by almost all organisms on the planet; gives off carbon dioxide as a waste gas.
- **Decomposition:** When plants and animals die, they are decomposed by bacteria and fungi. They also respire and give off carbon dioxide.
- **Combustion:** When organic material, such as wood, oil or coal, is burned, carbon dioxide is given off into the atmosphere.

Global effects in the atmosphere

Carbon dioxide in the atmosphere keeps the Earth warm by trapping heat from the sun as infrared rays – the **greenhouse effect**. Currently, there is a large increase in the amount of carbon dioxide in the atmosphere, which is making the Earth's global temperature increase. This is called **global warming**.

Effects of global warming include:

- sea levels rise
- different weather patterns
- increased flooding, drought and desertification
- migration of organisms.

Worked example

Why has the amount of carbon dioxide in the atmosphere increased recently? **(3 marks)**

Increased combustion of fossil fuels. More organisms on the planet leads to an increased level of aerobic respiration. Deforestation leads to decreased level of photosynthesis.

Now try this

🖩 Maths skills

1 In 1800, the carbon dioxide level in the atmosphere was 180 ppm. Today it is 400 ppm. What is the percentage increase in carbon dioxide? **(2 marks)**

2 Why is carbon important for our bodies? **(4 marks)**

Primary succession

Succession is a directional change in a community over time.

The process of primary succession

A community is all of the populations of different species living in an ecosystem. The species and number of each species change as the environment changes over time: from bare rock, to **pioneer species** and eventually the **climax community**.

The diagram shows how the ecosystem changes over hundreds of years.

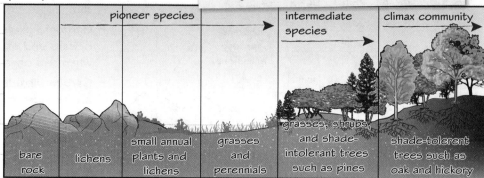

primary succession

pioneer species → intermediate species → climax community

bare rock | lichens | small annual plants and lichens | grasses and perennials | grasses, shrubs, and shade-intolerant trees such as pines | shade-tolerant trees such as oak and hickory

hundreds of years

- Lichens and mosses are common pioneer species. They break up the bare rock and when they die add a thin soil to the surface. Small plants and grasses can grow on this soil and add humus to the soil layer when they die.
- Over time, the soil is rich and deep enough for **intermediate species**, such as shrubs, to grow.
- Eventually, trees will be able to take root. This is described as the climax community, and is the most **stable**, with the highest level of **biodiversity**.

Each stage of succession is called a **sere**. In some ecosystems, such as sand dunes, all of the stages of succession can be seen at once. The newer sand dunes have pioneer species on them, whereas the older sand dunes have a climax community.

Deflected succession

If succession is prevented from running its course, this is known as **deflected succession**. This happens when humans interfere with the ecosystem, for example weeding a garden, or animals grazing a field.

Examples of succession

	Glacier	Sand dune
Pioneer species	Mosses Dwarf willows Mountain avens Willow-herbs	Sea rocket Prickly sandwort Sea couch grass Sea sandwort
Intermediate species	Alders	Marram grass Sea spurge
Climax species	Hemlock Sitka spruce	Hare's foot clover Bird's foot trefoil

Worked example

What adaptations would you expect a pioneer species to have? **(2 marks)**

The ability to tolerate extreme conditions and a mutualistic relationship with nitrogen-fixing bacteria in their root nodules

Pioneer plants would have to tolerate low nutrient levels. On a sand dune, there would also be salt spray and a lack of fresh water.

Mountain avens, hare's foot clover and bird's foot trefoil all have root nodules containing nitrogen-fixing bacteria.

Now try this

1 Why is the biodiversity of the climax community higher than the pioneer community? **(3 marks)**

2 Explain what would happen to a climax community if the soil became waterlogged. **(3 marks)**

Sampling

Sampling is a way of estimating the number of individuals of different species in a habitat. Look at page 80, to remind yourself of other sampling methods.

Measuring the distribution and abundance of organisms in heathland

When the rock or soil changes in an area, resulting in a change in the type of organisms living there, it is best to use a line or a belt transect to record the different species and estimate the number of each species. Transects can be a **continuous transect**, which records along the whole length of the transect, or an **interrupted transect**, which records at intervals along the tape. Here, a **belt transect** has been used to measure the abundance of each plant every 2 metres along the transect.

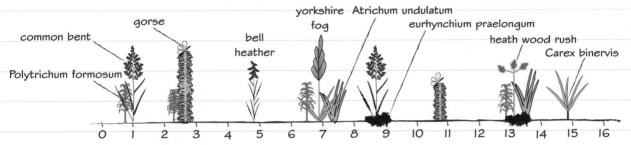

The **abiotic** factors along the transect should also be recorded. Now we can see the distribution of each plant, and can use the data from the abiotic and **biotic** factors to work out why.

🖩 Maths skills — Measuring the distribution and abundance of organisms in a forest

If the abundance and distribution of animals is being measured, it is best to use a method known as **capture, release, recapture**. The animals are caught in a pitfall trap, net or humane trap, and marked or tagged. Care must be taken to ensure that the method of marking does not decrease their chances of survival. After a period of time, animals should be trapped in the same area. The number of animals caught that do not have a mark and the number of animals caught that do have a mark should be recorded. The abundance of each species can be estimated using the equation:

$$\text{Population estimate} = \text{total first caught and marked} \times \frac{\text{total caught second time}}{\text{number of marked organisms in second catch}}$$

This method of estimating abundance assumes:

- Marked and unmarked animals have an equal chance of being captured.
- Marked animals have had enough time to intermix with the rest of the population.
- There is no emigration or immigration of animals.
- The population is stable (no births or deaths) in the time between the first catch and second catch.

Worked example

🖩 Maths skills

Estimate the population abundance if 84 mice are caught and tagged the first time, and then 96 are caught the second time and 37 of these had been tagged in the first capture. **(3 marks)**

$$\text{population} = 84 \times \frac{96}{37} = 218$$

Now try this

What are the advantages of using a continuous line transect compared with using quadrats when measuring the abundance of organisms on a sand dune? **(3 marks)**

You can check your answer:

- The first count selects 84 out of 218 mice = 0.385, so 38.5% of mice are tagged.
- The second count catches 37 out of 96 mice, which is also 38.5% of the mice.

Population sizes

Population sizes are controlled by limiting factors and follow a particular pattern.

Limiting factors

These are factors that determine the size of a population:

- food availability
- water availability
- **intraspecific competition** (competition between individuals of the same species)
- **interspecific competition** (competition between species)
- predation
- disease.

When a species moves into an area, the population growth always follows the same pattern: slow growth (lag phase), fast growth (log phase), and a fluctuating wave along the **carrying capacity**. This is the maximum number of a population that can survive in a habitat, due to the limiting factors.

Carrying capacity

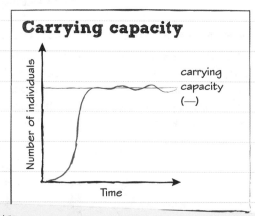

When the number in the population is higher than the carrying capacity (red line), some individuals will not survive. This brings the number in the population below the carrying capacity. The number in the population fluctuates around the carrying capacity.

Predator-prey relationships

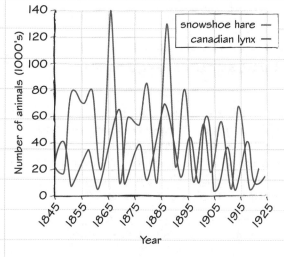

Predators and prey usually follow the same pattern or cycle. There are usually fewer predators than prey. As the number of prey increase, the number of predators increase, after a slight delay. This decreases the number of prey and, after a delay, decreases the number of predators.

Worked example

Why do the number of lynx increase after the number of snowshoe hares increase? **(3 marks)**

Food availability is a limiting factor on population size.

Lynx prey on the snowshoe hares / increased food availability.

Delay is due to time taken for lynx to breed.

Various factors may cause the snowshoe hare population to increase, such as increased food availability, or decrease, such as a reduction in breeding success when hare numbers are high. These will both have an impact on the lynx population.

Now try this

1 Which biotic factors would affect the population size of rabbits in a warren? **(3 marks)**
2 In the lynx/snowshoe hare example, explain what would happen if wolves move into the same area as the lynx. **(4 marks)**

Conservation

Conservation is important in order to protect species at risk as a result of human activity. You can look back at page 83 also, for more about maintaining biodiversity.

Conservation and preservation

- **Preservation** is the attempt to maintain a habitat in its current condition, without human intervention. A natural wilderness would be an example of a preserved habitat.

- **Conservation** is the active and dynamic management of a habitat, to allow the species that live there to thrive. It includes preventing succession, creating new habitats and protecting species from hunters. A national park or nature reserve would be examples of conserved habitats.

Reasons for conservation

	Reasons for conservation
Economical	Fishing Timber Tourism
Ecological	Pollination Water cleaning Nutrient cycling
Aesthetic	Personal well-being Holiday
Ethical	Right to life

Sustainability

If resources are taken away from a habitat and not replaced, eventually those resources will run out. If a habitat is managed in a **sustainable** way then resources are regularly replaced, so that they will not run out.

For example, for sustainable fishing, only a certain number of each fish species are allowed to be caught, to ensure a breeding population to provide fish for the future.

Sustainable timber production

1 Coppicing – tree trunk is cut and many smaller trunks grow from the stump.

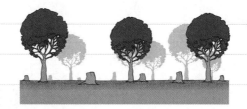

2 Selective felling – some trees in an area are cut down and replaced.

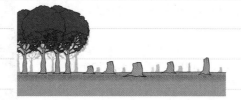

3 Clear felling – all trees in one area are cut down and replaced.

Worked example

How is conservation an active and dynamic process?

(4 marks)

active protection of species from poachers

captive breeding of endangered species

prevention of succession by managing the habitat

national and international laws preventing pollution of habitats / trade in endangered species

Areas of land may be fenced off and rangers employed to protect species and habitats.

Conservation in zoos and botanical gardens often includes captive breeding.

Grazing animals or burning heath will prevent succession.

Look back at page 83 for more details on, for example, CITES and the Rio convention on biological diversity.

Now try this

1 What are the differences between conservation and preservation? **(3 marks)**

2 What would happen to a habitat if timber production were not sustainable? **(4 marks)**

Managing an ecosystem

Endangered animals and habitats must be protected. But this has to be balanced with the need for humans to live on the same land, by managing the ecosystem.

The management of environmental resources

This can be done by creating conservation areas, and allowing tourists into the area to provide an income for the people who live there.

For example, in the Masai Mara region in Kenya, the Masai communities own all of the land between the national parks, including wildlife protection corridors and habitat reserves. Many programmes have been set up that promote sustainable economic benefits from preserving and conserving the ecosystem, including ecotourism, conservation and wildlife monitoring.

Other ecosystems that are actively managed in this way include:

- Antarctica
- the Terai region of Nepal
- peat bogs
- Snowdonia National Park
- the Lake District.

Issues

It is often difficult to preserve a habitat while balancing the human needs for housing, growing food and making money from the land, for example hunting. Conservation techniques often cause less conflict, as they allow a controlled amount of human interaction, such as hunting quotas.

Effects of human activities

Human activity has a huge impact on the environment; for example, pollution, building, and using land for farming.

In **ecosensitive** ecosystems, such as the Galapagos Islands, human activity has to be monitored closely in order to conserve the **endemic** species.

Problems arising from human activity in the Galapagos:

- invasive plants
- invasive insects
- disease
- over-fishing
- introduction of pets.

The Galapagos Marine Reserve was set up in 1986 to protect wildlife on the islands. There are also breeding programmes on the islands to increase the populations of the iguanas and giant tortoises.

Worked example

Explain why some of the islands in the Galapagos have strictly limited visitation.   **(4 marks)**

preservation of habitat

prevent invasive plant / insect species

prevent disease spreading

decrease pollution

Now try this

1 Explain why there is conflict between tourism and conservation in the Masai Mara.   **(3 marks)**

2 Suggest what the effect would be of an invasive insect species on the Galapagos Islands.

    (3 marks)

Exam skills

These questions use knowledge and skills that you have already revised. Look at pages 166–168 to revise populations, conservation and managing ecosystems.

Worked example

(a) Name two limiting factors that determine the size of a population. **(2 marks)**

food / water availability

competition

predation / disease

Competition can be intraspecific (between individuals of the same species) or interspecific (between individuals of different species).

(b) What is the carrying capacity of a population? **(1 mark)**

This is the maximum number of a population that can survive in a habitat, due to the limiting factors.

The number of individuals living in a habitat increases until it reaches the carrying capacity. The number in the population then oscillates around carrying capacity.

(c) What are some of the economical and ecological reasons for conservation? **(2 marks)**

Economical – can make money from fishing, timber or tourism.

Ecological – The natural world maintains the water cycle and the nitrogen cycle.

You could also include pollination as an ecological reason for conservation.

(d) What is the difference between conservation and preservation? **(2 marks)**

Preservation is the attempt to maintain a habitat in its current condition, without human intervention. Conservation is the active and dynamic management of a habitat, to allow the species that live there to thrive.

A natural wilderness is an example of preservation. A national park is an example of conservation.

(e) Why do some ecosystems have to be managed? **(2 marks)**

There is often conflict between the need for endangered animals and habitats to be protected and the need for humans to live on the same land.

For example, people make money by digging up peat from peat bogs, but the removal of the peat destroys the habitat.

(f) Explain what problems human activity causes on the Galapagos Islands. **(4 marks)**

Humans may bring invasive plants or insects to the islands, which do not have any natural predators. Humans may also bring animals to the islands that compete with the native species. Visitation to the islands may bring diseases that infect the native species, or increase air or water pollution.

There are many endemic species on the Galapagos Islands that could be lost if a pest or disease was introduced to the islands.

Over-fishing is also a problem caused by humans.

Answers

These answers provide key points. In some cases, other approaches may also be suitable. In your exam you should aim to write in full sentences.

Module 2

1. Using a light microscope

1   ×200 **(1)**
2   Observe in colour. **(1)**
    Observe the cells in real time. **(1)**

2. Using other microscopes

Type of microscope	Can observe whole cells and tissues	Can observe organelles	Can observe cell surfaces	Can observe a certain depth within a cell	
Transmission electron microscope	✓	✓			**(1)**
Scanning electron microscope	✓		✓		**(1)**
Laser scanning confocal microscope	✓	✓	✓	✓	**(1)**

3. Preparing microscope slides

1   $10 \times 10\,\mu m = 100\,\mu m$ **(1)**
    $100\,\mu m \backslash 100 = 1\,\mu m$ **(1)**
2   Different blood cells / red blood cells (erythrocytes) and white blood cells (leukocytes) **(1)**
    Nucleus of the white blood cell (leukocyte) **(1)**

4. Calculating magnification

1   (a) $12\,cm = 120\,000\,\mu m$ **(1)**
    (b) $16\,mm = 16\,000\,\mu m$ **(1)**
2   $33\,mm = 33\,000\,\mu m$ **(1)**
    $\dfrac{33\,000}{15\,000} = 2.2\,\mu m$ **(1)**

5. Eukaryotic and prokaryotic cells

1   Both prokaryotic and eukaryotic cells can move using flagella (undilopodium in eukaryotes). **(1)**
    Both prokaryotic and eukaryotic cells can move using cilia. **(1)**
2   Rough endoplasmic reticulum / ER has a rough appearance due to presence of ribosomes. **(1)**
    Smooth endoplasmic reticulum / ER has a smooth appearance. **(1)**
    Rough ER is the site of protein synthesis. **(1)**
    Smooth ER is the site of lipid synthesis. **(1)**

6. The secretion of proteins

1   The rough ER contains ribosomes. **(1)**
    Ribosomes translate the messenger RNA / mRNA. **(1)**
    Rough ER packages proteins into vesicles. **(1)**
2   Plasma cells contain more rough ER and Golgi bodies than other cells. **(1)**
    Ribosomes produce antibodies. **(1)**
    Golgi body modifies antibodies. **(1)**
    Golgi body packages modified antibodies into vesicles **(1)** for exocytosis. **(1)**   **(max 4)**

7. The cytoskeleton

1   The cytoskeleton is made of many microtubules. **(1)**
    Proteins called microtubule motors move up and down the microtubules. **(1)**
    Proteins in vesicles are carried by the microtubule motors. **(1)**
2   Cilia are found in the oviducts / Fallopian tubes **(1)**
    where they move / waft the ovum / egg towards the uterus. **(1)**
    Spermatozoa / sperm cells have undulipodia **(1)**
    which they use to swim towards the ovum / egg. **(1)**

8. The properties of water

1   Water is a polar molecule. **(1)**
    cohesion between water molecules and solute due to hydrogen bonds forming **(1)**
    adhesion between water molecules and wall / xylem / blood vessel due to hydrogen bonds forming **(1)**
2   Water has a high latent heat of vaporisation. / large amount of energy required to change from liquid to gas. **(1)**
    Water evaporates from surface, taking heat with it, **(1)** for example, sweating / panting / transpiration. **(1)**

9. The biochemistry of life

1   (b) $CH_3(CH_2)_{10}CO_2H$ **(1)**
2   iron ion / Fe^{2+} in haemoglobin **(1)**
    zinc ion / Zn^{2+} in DNA polymerase **(1)**
    magnesium ion / Mg^{2+} in chlorophyll **(1)**   **(max 2)**

10. Glucose

1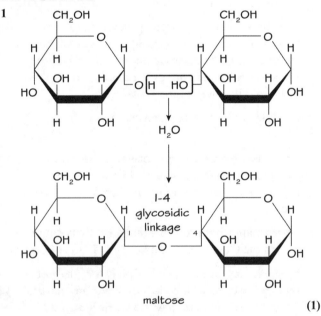

maltose **(1)**

2   Glycogen contains α-glucose and forms α-1–4-glycosidic bonds. **(1)**
    This causes the glycogen chain to coil. **(1)**
    Glycogen also branches using α-1–6-glycosidic bonds. **(1)**
    Cellulose contains β-glucose and forms β-1–4-glycosidic bonds. **(1)**
    This causes the cellulose to form straight chains. **(1) (max 3)**

11. Starch, glycogen and cellulose

Cellulose is made of long straight chains of β-glucose **(1)**
connected by hydrogen bonds. **(1)**
Used in cell walls to give them strength. **(1)**

12. Triglycerides and phospholipids

1 Triglycerides contain one glycerol and three fatty acids. **(1)**
Phospholipids contain one glycerol, two fatty acids and a phosphate group. **(1)**

2

triglyceride   glycerol   fatty acids

13. Uses of lipids in living organisms

1 energy storage **(1)**
insulation **(1)**
buoyancy **(1)**
prevents membranes from freezing **(1)**   **(max 3)**

2 Steroid hormones are non-polar. **(1)**
Can pass through plasma membrane without need for intrinsic / carrier / channel protein. **(1)**

14. Amino acids

1 an amino group on one side of the central carbon **(1)**
a carboxyl group on the other side of the central carbon **(1)**
an R group below the central carbon **(1)**

2 Amino acids contain oxygen and hydrogen atoms. **(1)**
A hydrogen bond forms between an oxygen atom on one amino acid and a hydrogen atom on another amino acid. **(1)**

15. Protein

1 quaternary structure **(1)**
2 in the centre of the protein **(1)**

16. Fibrous proteins

1 Quaternary proteins have two or more polypeptide chains. **(1)**
Collagen is made of three polypeptide chains. **(1)**

2 three polypeptides wound around each other **(1)**
cross-links form between the polypeptides **(1)**
forms collagen fibrils and collagen fibres. **(1)**

17. Globular proteins

1 A conjugated protein is a protein that requires a non-amino group in order to function. **(1)**
The non-amino groups is attached to the protein. **(1)**

2 Disulfide bridges form between cysteine amino acids. **(1)**
Haemoglobin has no cysteine amino acids. **(1)**

18. Benedict's test

1 to break the glycosidic bond (between the glucose and the fructose) **(1)**
2 The sugar is oxidised. **(1)**

19. Tests for protein, starch and lipids

1 Milk contains lipids. **(1)**
A white emulsion will form **(1)**
after dissolving in ethanol and adding to water. **(1)**   **(max 2)**

2 The sample contains starch. **(1)**
The enzyme digests starch to maltose. **(1)**
Maltose is a reducing sugar. **(1)**
Maltose will reduce Benedict's reagent. **(1)**   **(max 3)**

20. Practical techniques – colorimetry

1 absorbance **(1)**
2 to detect the light **(1)**
that passes through the sample **(1)**

21. Practical techniques – chromatography

1 0.78 **(1)**
2 $12 \times 0.4 = 4.8 \, \text{cm}$ **(2)**
3 Different components will have a different attraction to the paper. **(1)**
Components with a high affinity bind strongly to the paper. **(1)**
Components with a low affinity travel far up the paper. **(1)**

23. Nucleotides

1 The deoxyribose sugar and phosphate groups **(1)**
joined by phosphodiester bonds. **(1)**
2 messenger RNA / mRNA **(1)**
transfer RNA / tRNA **(1)**
ribosomal RNA / rRNA **(1)**

24. ADP and ATP

hydrolysis **(1)**

25. The structure of DNA

1 TACAAAGATTGGTTCCCG **(1)**
2 It is made up of alternating deoxyribose sugars **(1)** and phosphate groups. **(1)**

26. Semi-conservative DNA replication

1 a change in the nucleotide base sequence in the DNA **(1)**
2 The new DNA always contains one strand of original DNA **(1)** and one strand of new DNA. **(1)**

27. The genetic code

Genes can be taken out of one organism and placed into another. **(2)**

28. Transcription and translation of genes

1 They all contain the RNA bases, A, C, G and U. **(1)**
They are all single stranded. **(1)**
2 mRNA is a long, single strand. **(1)**
rRNA forms the ribosome. **(1)**
tRNA carries an amino acid. **(1)**

29. The role of enzymes

1 (a) breaks down sucrose **(1)**
(b) breaks down dipeptides **(1)**
(c) breaks down lipids **(1)**
2 Enzymes speed up the rate of reaction. **(1)**
lower the activation energy needed to carry out the reaction **(1)**
lower the temperature needed to carry out the reaction **(1)**
do not get used up by the reaction **(1)**
can work at body temperature / normal pressure / neutral pH. **(1)**   **(max 3)**

30. The mechanism of enzyme action

1 The enzyme is specific to its substrate in both models. **(1)**
In the induced-fit hypothesis, the active site moulds around the substrate, after the substrate has bound. **(1)**
2 Research into enzyme shape, before and after the substrate **(1)** binds, shows that the enzyme changes shape after binding. **(1)**

31. Factors that affect enzyme action

1 Pepsin's optimum pH is 2. **(1)**
The increase in pH will break the hydrogen bonds. **(1)**
The tertiary structure of the active site will denature. **(1)**
The substrate will no longer be able to bind to pepsin. **(1)**   **(max 3)**

2 As the enzyme concentration increases, the rate of reaction increases. **(1)**
This is because the substrate is more likely to collide with the enzyme. **(1)**

32. Factors that affect enzyme action – practical investigations

1 Do not include anomalous result in the mean. **(1)**
Repeat the investigation. **(1)**
2 Some of the oxygen gas may not be measured **(1)**
due to the oxygen escaping from the apparatus / oxygen dissolving in the water. **(1)**
or
difficulty in starting the timer as hydrogen peroxide is added to catalase **(1)**
More oxygen given off by reaction than measured in 5 minutes. **(1)**

33. Cofactors, coenzymes and prosthetic groups

Same charges repel / opposite charges attract. **(1)**
Altering the charge prevents substrate being repelled by active site. **(1)**

34. Inhibitors

1 Cyanide **(1)**
non-competitive inhibitor / binds to allosteric site **(1)** inhibits cytochrome oxidase, used in aerobic respiration **(1)**
or
Ethylene glycol **(1)**
competitive inhibitor / binds to the active site **(1)** inhibits pyruvate dehydrogenase, used in respiration. **(1)**
2 ACE inhibitors **(1)**
inhibit angiotensin converting enzyme / ACE **(1)**
prevent ACE from increasing blood pressure **(1)**
or
ATPase inhibitor **(1)**
used to treat heart problems / arrhythmia **(1)**
binds to sodium–potassium pumps / allows more calcium ions into
the heart / increases the heart rate. **(1)**

36. The fluid mosaic model

separate the contents of the mitochondria from the cytoplasm **(1)**
control the movement of molecules in and out of the mitochondria **(1)**

37. Factors that affect membrane structure

1 Cholesterol is embedded within the phospholipid bilayer. **(1)**
Cholesterol prevents phospholipids packing too close together. **(1)**
2 Ethanol dissolves lipids. **(1)**
Plasma membranes are made of phospholipids. **(1)**
The phospholipids in the plasma membranes of beetroot would be damaged. **(1)**

38. Movement across the membrane

1 The plasma membrane is made out of phospholipids. **(1)**
Lipids are non-polar. **(1)**
2 Oxygen is needed for aerobic respiration to make ATP. **(1)**
Active transport needs ATP. **(1)**
Active transport will slow or stop. **(1)**

39. Osmosis

The hypotonic solution has a lower / more negative water potential than inside the red blood cell. **(1)**
Water will move out of the red blood cell from high water potential to low water potential. **(1)**
The red blood cell will shrink / crenate. **(1)**

40. Movement across the membrane – practical investigations

$0.28\,mol/dm^3$ **(2)**

41. The cell cycle

1 Stem cells / tumour cells have no Hayflick limit. **(1)** They can go through the cell cycle an infinite number of times. **(1)**
2 Checkpoint proteins check the cell for damaged DNA / mutations. **(1)**
If damage / mutations are detected, the cell is prevented from going to the next phase. **(1)**

42. Mitosis

Duplicated chromosomes are supercoiled and become visible. **(1)**
Chromosomes line up on the equator of the cell. **(1)**
Each sister chromatid is separated by spindle fibres. **(1)**
A set of chromosomes gathers at each end of the cell. **(1)**

43. Meiosis

random assortment **(1)**
crossing over **(1)**

44. Specialised cells

Erythrocytes have no nucleus, neutrophils have a lobed nucleus. **(1)**
Erythrocytes are biconcave, neutrophils can change shape. **(1)**
Erythrocytes are packed with haemoglobin and neutrophils are packed with lysosomes. **(1)**

45. Specialised tissues

Xylem is hollow to allow water to travel up the plant. **(1)**
Xylem is strong due to the lignin. **(1)**
Xylem is waterproof which keeps the water inside the xylem. **(1)**

46. Stem cells

1 A stem cell is an undifferentiated cell. **(1)**
It can differentiate into any type of cell for that organism. **(1)**
2 Stem cells in the bone marrow are multipotent **(1)**
develop into any type of blood cell / haematopoietic **(1)**
differentiate into neutrophils. **(1)**

47. Uses of stem cells

Adult stem cells extracted from patient. **(1)**
Cells stimulated to differentiate into a particular cell type. **(1)**
Cells grown over a scaffold to achieve a particular shape. **(1)**

Module 3

49. Gas exchange surfaces

Gases diffuse from high concentration to low concentration. **(1)**
The steeper the concentration gradient, the faster the rate of diffusion. **(1)**

50. The lungs

1 Mucus builds up inside the lungs / difficult to move mucus **(1)**
heavy, persistent cough / smoker's cough **(1)**
2 Asthma is an overreaction of the lungs to substances in the air. **(1)**
Smooth muscle contracts. **(1)**
This constricts the airway / narrows the lumen. **(1)**

51. The mechanism of ventilation

1 Air enters the lungs very quickly **(1)**
   and is stopped by the glottis closing, making a hiccup sound. **(1)**
2 less atmospheric pressure **(1)**
   less difference between high and low pressure inside and outside of the lungs **(1)**
   Intercostal and diaphragm muscles have to work harder in order to get air into the lungs. **(1)**

52. Using a spirometer

$$\left(\frac{1.8}{45}\right) \times 60 \ \textbf{(1)}$$

= 2.4 dm³ per minute **(1)**

53. Ventilation in bony fish and insects

1 The blood flows in the opposite direction to the water. **(1)**
   Water is highly oxygenated compared to blood at all points along the lamellae **(1)**
   maintains a high concentration gradient **(1)**
   absorbs the maximum amount of oxygen. **(1)** **(max 3)**
2 Movement of wing muscles increases the volume of thorax. **(1)**
   Moving the tracheal fluid into the cells increases the surface area of the tracheae for oxygen diffusion. **(1)**

54. Circulatory systems

Blood flows faster in a closed circulatory system. **(1)**
Blood can be diverted in a closed circulatory system. **(1)**

55. Blood vessels

1 arteries have a thicker wall, veins have a thinner wall / arteries have a thicker layer of smooth muscle / thicker layer of collagen fibres. **(1)**
   Arteries do not have valves, veins have valves. **(1)**
   Arteries have a narrow lumen, veins have a wide lumen. **(1)**
2 small lumen **(1)**
   recoil of elastic fibres **(1)**

56. The formation of tissue fluid

Blood plasma contains large proteins. **(1)**
Blood plasma contains more dissolved substances / oxygen / glucose / amino acids. **(1)**

57. The mammalian heart

Blood in pulmonary artery is deoxygenated / blood in aorta is oxygenated. **(1)**
Both under pressure, but blood in aorta is at a higher pressure than the blood in the pulmonary artery. **(1)**

58. The cardiac cycle

1 The line increases slightly up to 5 mmHg and then decreases. **(1)**
   The line then increases steeply up to 120 mmHg and then decreases steeply. **(1)**
   The line decreases gradually down to 0 mmHg. **(1)**
2 The semilunar valve will open. **(1)**
   Blood can flow into the aorta through the semilunar valve. **(1)**

59. Control of the heart

1 to make sure that the blood flows upwards towards the pulmonary artery and aorta **(1)**
   to fill the valve pockets of the atrioventricular valves and cause them to close, preventing the backflow of blood **(1)**
2 same size P wave, QRS complex and T wave **(1)**
   different intervals between end of T wave and beginning of next P wave **(1)**

60. Haemoglobin

1 four molecules of oxygen **(1)**
2 More oxygen is released from the haemoglobin **(1)**
   to be used in aerobic respiration. **(1)**

61. The plant vascular system

1 to provide support for the phloem sieve tube elements **(1)**
   provide ATP for the active transport of sugars in and out of the phloem **(1)**
2 The wall of the xylem contains lignin. **(1)**
   Xylem has no end plates between cells but phloem has sieve plates between its cells. **(1)**
   The xylem wall contains pits. **(1)**

62. Leaves, stems and roots

1 Xylem is a continuous tube. **(1)**
   Lignin forms rings in the xylem wall. **(1)**
2 contains meristem tissue / stem cells **(1)**
   makes new xylem and phloem **(1)**

63. Transpiration

1 Photosynthesis happens in daylight **(1)**
   needs carbon dioxide from the air which enters through the stomata. **(1)**
2 Prepare plant and potometer under water to prevent air bubbles in the xylem. **(1)**
   Shine a lamp onto the plant at different distances. **(1)**
   Introduce a bubble to the capillary tube. **(1)**
   Observe the movement of the bubble every 3 minutes. **(1)**
   Repeat the experiment. **(1)**
   Use a water tank to prevent heat from the lamp affecting the results of the experiment. **(1)** **(max 5)**

64. The transport of water

Coiled leaves and sunken stomata **(1)**
prevent wind removing water vapour. **(1)**
Hairs on the underside of the leaf **(1)**
trap water droplets. **(1)**
These decrease the water vapour potential gradient. **(1)** **(max 5)**

65. Translocation

1 A source is where the sugars come from. **(1)**
   A sink is where the sugars are going to. **(1)**
2 In the experiment, water moves into a concentrated solution **(1)**
   creating a pressure **(1)** which moves the solution through a tube. **(1)**

Module 4

67. Types of pathogens

1 A bacterium is a prokaryote and a virus is not a true organism. **(1)**
   Bacteria are much larger than viruses. **(1)**
   Bacteria have a cell wall and viruses have a protein coat. **(1)**
   Bacteria have DNA in chromosomes and plasmids. **(1)**
2 Malaria is caused by a protoctista that is carried in the saliva of mosquitos. **(1)**
   Mosquitos spread the protoctista when they bite humans to feed on their blood. **(1)**

68. Transmission of pathogens

1 spores carried by the wind **(1)**
   spores carried by a vector / beetle / *Scolytus multistriatus* **(1)**
2 Warm environments more likely to have vectors / mosquitos. **(1)**
   Cramped environments will spread diseases by direct contact / touching / coughing / sneezing. **(1)**

69. Plant defences against pathogens

1 Stomata close when the plant detects any bacteria or fungi. **(1)**
2 Ring rot is caused by a bacterium. **(1)**
  Close stomata. **(1)**
  Callose formation in the phloem. **(1)**   **(max 2)**

70. Animal defences against pathogens

1 hydrochloric acid in the stomach **(1)**
  expulsion from the body / vomiting **(1)**
2 Antihistamine relieves symptoms of hay fever **(1)**
  prevents expulsive reflexes / sneezing / eyes watering. **(1)**

71. Phagocytes

Macrophages are one type of antigen-presenting cell **(1)**
present pathogen's antibodies on their surface. **(1)**
Antigens stimulate T lymphocytes / specific immune response. **(1)**

72. Lymphocytes

1 Antibodies are produced by the ribosomes / rough ER. **(1)**
  Golgi body modifies antibodies for export from the cell. **(1)**
  Antibodies travel to the plasma membrane for exocytosis in vesicles along the cytoskeleton. **(1)**
2 Cells infected with virus display viral antigen on cell surface. **(1)**
  Killer T cells recognise the viral antigen. **(1)**
  Killer T cells destroy the cell, using cytotoxins that trigger apoptosis / programmed cell death, and the virus inside it. **(1)**

73. Immune responses

Agglutinins make the pathogens clump together. **(1)**
Pathogens can no longer enter cells. **(1)**
Pathogens are marked for phagocytosis. **(1)**

74. Types of immunity

1 active **(1)**
  artificial **(1)**
2 In active immunity you make the antibodies, not just given. **(1)**
  You can make the same antibodies again. **(1)**
  Not dependent on an external source of antibodies. **(1)**

75. The principles of vaccination

1 A toxoid is a harmless version of a toxin. **(1)**
  When injected, antibodies are made that bind to the toxin. **(1)**
  When the body is exposed to the toxin, the antibodies bind to it and prevent the toxin from harming the body. **(1)**
2 no longer have herd immunity **(1)**
  people who cannot be vaccinated no longer protected **(1)**
  increased incidence of disease / measles / whooping cough **(1)**

76. Vaccination programmes

1 the rapid spread of an infectious disease **(1)**
2 Influenza virus mutates every year **(1)**
  different antigen on surface of virus **(1)**
  need new vaccine to stimulate antibody production against the new antigen. **(1)**

77. Antibiotics

1 New medicines are often found in plants. **(1)**
  There may be undiscovered medicines in threatened areas. **(1)**
2 people not taking the full course of antibiotics **(1)**
  uncontrolled use of antibiotics **(1)**

79. Measuring biodiversity

1 Species richness is the number of different species. **(1)**
  Species evenness is the number of individuals of each species. **(1)**
  Species richness is more important for biodiversity because a wide variety of species leads to an increase in genetic variation. **(1)**

2 Random sampling uses random numbers to place quadrats in a certain area. **(1)**
  Removes bias. **(1)**
  Non-random sampling can be opportunistic. **(1)**
  Non-random sampling can be stratified. **(1)**
  Transects are a form of non-random sampling / systematic. **(1)**   **(max 4)**

80. Sampling methods

Sweep net **(1)**

81. Simpson's index of diversity

1 Low biodiversity means fewer species or fewer individuals of each species **(1)**
  less genetic variation, so less able to adapt to new selective pressures **(1)**
  fewer food sources, so species may be less likely to survive if one food source is reduced. **(1)**
2 $n/N = 0.2 / 0.2 / 0.1 / 0.2 / 0.1$ **(1)**
  $(n/N)^2 = 0.04 / 0.04 / 0.01 / 0.04 / 0.01$ **(1)**
  $D = 0.86$ **(1)**

82. Factors affecting biodiversity

1 $7500 \backslash 30\,000$ **(1)**
  $= 0.25$ **(1)**
2 land lost to desertification **(1)**
  water levels rising **(1)**
  less water availability **(1)**
  species migration **(1)**

83. Maintaining biodiversity

1 Genetic variation will be decreased **(1)**
  due to decreased population size. **(1)**
  Seeds collected might not be representative of all alleles in a species. **(1)**
2 conserves endangered / rare species **(1)**
  increased cooperation between countries **(1)**
  difficult to stop all illegal trade **(1)**
  Not all countries are signed up to the treaties. **(1)**

84. Classification

1 (a) Canis **(1)**
  (b) Mustelidae **(1)**
2 Organisms with similar physical features are grouped together **(1)** into gradually smaller categories. **(1)**

85. The five kingdoms

1 Protoctista **(1)**
  plus explanation
  It is autotrophic **(1)**
  cannot be a plant as there are no cell walls. **(1)**   **(max 2)**
2 Classification places species into groups. **(1)**
  Phylogeny places species on a continuum. **(1)**
  Classification uses anatomy / physical characteristics. **(1)**
  Phylogeny uses DNA / biochemistry. **(1)**
3 The DNA is extracted from species. **(1)**
  The genetic code is sequenced. **(1)**
  The DNA sequence from each species is compared. **(1)**

86. Types of variation

lower than the critical value at $p = 0.05$ **(1)**
would accept the null hypothesis **(1)**
There is greater than 5% probability that these results occurred by chance. **(1)**

87. Evolution by natural selection

1 wide feet for walking on sand **(1)**
  large ears to radiate heat **(1)**
  long eyelashes to keep out sand **(1)**
  sand coloured / camouflage **(1)**

ability to store fat (food) in such a way that it doesn't provide heat insulation (camel's hump) **(1)**

nocturnal **(1)** **(max 3)**

2   Green individuals are camouflaged / escape predation. **(1)** Blue individuals less likely to escape predation. **(1)** Green individuals more like to have offspring. **(1)** More of population will be green. **(1)**

88. Evidence for evolution

1   species C and D **(1)**
and
They have the most similar DNA sequences. **(1)** Only one base is different. **(1)** **(max 2)**

2   can only see hard tissue / bone / shell **(1)** unless organism has been preserved in peat / ice **(1)** Intermediates missing from the fossil record. **(1)**

Module 5

90. The need for communication

1   Effector cells are often some distance from receptor cells **(1)**; response needs to be targeted **(1)**; response sometimes needs to be quick **(1)**; internal conditions need to be maintained. **(1)** **(max 3)**

2   temperature **(1)**; pH **(1)**; water balance **(1)**; freedom from toxins **(1)**; inhibitor maintenance **(1)** **(max 3)**

91. Principles of homeostasis

1   negative feedback **(1)**

2   enzymes **(1)**; would begin to denature **(1)**; metabolic reactions would stop working **(1)**; suitable consequence, e.g. death **(1)** **(max 3)**

3   The skin contains peripheral temperature receptors **(1)**; but also has a role in maintaining overall body temperature, e.g. sweating, vasoconstriction/dilation, hairs. **(1)**

92. Temperature control in endotherms

1   allow early detection of temperature change **(1)**; protect the inner body/organs from temperature changes **(1)**; allow a steady core temperature to be maintained **(1)** **(max 2)**

2   to share warmth **(1)**; reduce the surface area exposed to the cold **(1)**

93. Temperature control in ectotherms

they move around little **(1)**; don't maintain body temperature **(1)**; little energy is required therefore they respire less **(1)**; have low metabolic requirements **(1)**

94. Excretion

lungs **(1)**; kidneys **(1)**; skin **(1)**

95. The liver – structure and function

no bile storage; bile constantly passes into small intestine. **(1)**

96. The kidney – structure and function

even higher pressure in the glomerulus **(1)**; capillaries may have burst **(1)**; blood cells not reabsorbed **(1)** **(max 2)**

97. Osmoregulation

1   Otherwise, water would constantly be reabsorbed from the collecting duct. **(1)**

2   It would take time **(1)**; and energy **(1)**; to break them down **(1)**; and re-synthesise them when required. **(1)** **(max 3)**

98. Kidney failure and urine testing

1   So that there is no risk that the blood will become infected (with bacteria for example) **(1)**; To match body temperature (when the blood returns to the body). **(1)**

2   Humans have two kidneys **(1)**; it is possible to survive (and live a 'normal' life) with one. **(1)**

100. Sensory receptors

converts stimuli into nerve impulses **(1)**; (a transducer is something which converts energy from one form into another) **(1)**

101. Types of neurone

1   It speeds it up. **(1)**

2   An axon occurs after the cell body **(1)**; in the direction of an impulse/action potential **(1)**; the dendron occurs before the cell body **(1)**; the dendron carries an action potential to the cell body from the sensory receptor **(1)**; the axon carries the action potential into the CNS. **(max 2)**

3   Yes. **(1)** energy/ATP **(1)**; protein synthesis **(1)**; active transport **(1)** **(max 2)**

102. Action potentials and impulse transmission

1   The arrival of **(1)**; an action potential, e.g. the movement of Na^+ into the cell **(1)**; changing the potential difference across the cell membrane. **(1)** **(max 2)**

2   The potential difference is greater than the resting potential **(1)**; less chance of reaching threshold potential **(1)**; an action potential is not produced **(1)**; some of the Na^+ are still in the cell which would reduce the rate of influx. **(1)** **(max 3)**

103. Structure and roles of synapses

1   energy/ATP **(1)**; active transport of acetyl/choline **(1)**; acetylcholine synthesis **(1)**; vesicle synthesis **(1)**; vesicle transport **(1)** **(max 2)**

2   short diffusion distance **(1)**; take too long if bigger **(1)**

3   Na^+ **(1)**, K^+ **(1)**, Ca^{2+} **(1)**

104. Endocrine communications

1   via the blood **(1)**

2   to transfer the message from the hormone **(1)**; to activate other molecules, e.g. enzymes **(1)**; inside the cell **(1)** **(max 2)**

105. Endocrine tissues

tertiary structure **(1)** changes shape/moves or alters position **(1)**

106. Regulation of blood glucose

1   in the islets of Langerhans **(1)**; in the pancreas **(1)**

2   a change in the potential difference across a cell membrane **(1)**; causes Ca^{2+} channels to open **(1)**; Ca^{2+} diffuse into the cell **(1)**; Ca^{2+} cause vesicles (of neurotransmitter or insulin) to fuse with the cell membrane. **(1)** **(max 3)**

107. Diabetes mellitus

1   no animals hurt or killed **(1)**; bacteria are not eukaryotic/are simple life forms. **(1)**

2   a factor which doesn't cause a disease by itself **(1)**; but can contribute to the onset of a disease **(1)**; type II diabetes occurs due to a number of different factors **(1)** **(max 2)**

108. Plant responses to the environment

1   allows gibberellin to reach the aleurone layer **(1)**; provides a medium/solvent for reactions to occur **(1)**

2   lack of cellulase production **(1)**; cellulose not broken down in abscission zone **(1)**

109. Controlling plant growth

1 The evidence shows this not to be true (1); auxin levels do not directly cause this, cytokinins promote lateral bud growth. (1)
2 cell division (1); cell elongation (1)

110. Plant responses

1 diffusion or facilitated diffusion (1)
2 All auxin might be destroyed during day or in bright light (1); plant would have to make more auxin (1); normal growth would be affected (1); energy would be used making more auxin. (1)　　　　　　　　　　　　　　(max 2)

111. Commercial use of plant hormones

1 Bananas produce ethene which help avocados to ripen. (1)
2 Broad leaved weeds absorb more (1); leaves grow too fast (1); roots/transport systems cannot support growth. (1)

112. Mammalian nervous system

1 When upregulated it prepares the body for action (1); increased ventilation (1); increased heart rate (1); more blood to muscles (1); and brain (1); less to the gut. (1)　(max 3)
2 self-governing (1)

113. The brain

1 medulla oblongata (1)
2 damages non-essential areas (1); areas which govern thought and emotion could be damaged (1)

115. Coordination of responses

Adrenaline (1); causes blood to be diverted away from the gut (1); vasoconstriction (1); and towards the brain and muscles. (1)
　　　　　　　　　　　　　　　　　　(max 2)

116. Controlling heart rate

1 It maintains the heart rate. (1)
2 Carbon dioxide reacts with water to produce carbonic acid (1); this dissociates to H^+ and HCO_3^- (1); causing a decrease in blood pH (1)　　　　　　　　　　(max 2)

117. Muscle structure and function

1 contains large quantities of actin (1); and myosin (1)
2 One does work in one direction (1); no force generated when relaxes (1); antagonistic muscles work against each other/in opposite directions. (1)

118. Muscular contraction

less muscle contraction (1); tropomyosin would remain covering the myosin binding sites (1); on actin filaments (1)
　　　　　　　　　　　　　　　　　　(max 2)

120. Photosynthesis and respiration

1 In the dark plants only respire (like animals – they only ever respire) (1); in the light plants respire (1); and photosynthesise. (1)
2 (a) It would lower the water potential of the chloroplast. (1)
　(b) The chloroplast could swell (and burst) (1); converting glucose to starch (1); starch would not affect the water potential of the chloroplast. (1)　(max 2)

121. Photosystems, pigments and thin layer chromatography

1 low light levels (1); have lots of pigment (1); different kinds of pigment (1)　　　　　　　　　　(max 2)
2 in photosystems (1)

122. Light-dependent stage

1 NADP is not needed for cyclic photophosphorylation (1); no electrons are lost from chlorophyll a in PSI. (1)
2 so gas exchange can take place (1); carbon dioxide is needed (for photosynthesis/the LIS) (1); so transpiration can occur. (1)　　　　　　　　　　　　　　(max 2)

123. Light-independent stage

1 They need CO_2 (1); to make cell walls (1); starch, proteins, etc. (1); photorespiration (1)　　　　　(max 2)
2 phosphorylation of GP (1); to help re-cycle TP to RuBP (1)
3 converted to GP (1); Calvin cycle intermediate (1); not used in the synthesis of other molecules (1)　　(max 2)

124. Factors affecting photosynthesis

1 Light is not needed for the enzyme RuBisCO to work (1); can continue as long as CO_2, ATP and reduced NADP are available. (1)
2 RuBP still gets converted to GP (1); less ATP for LDS (1); less TP re-cycled to RuBP. (1)　　　　(max 2)

125. The need for cellular respiration

1 releases small amounts of energy (1); locally synthesised (1); re-cyclable (1)　　　　　　　　　(max 2)
2 photoautotrophs (1)

126. Glycolysis

1 4 ATP are made (1); but 2 are used in activation at the start (1)
2 NAD (1); triose phosphate (1)
3 cytosol/cytoplasm (1)

127. Structure of the mitochondrion

1 matrix (1)
2 provide a large surface area (1); for ATP synthase (1); electron transport chain proteins (1) (max 2)
3 DNA (1); and ribosomes (1)

128. Link reaction and the Krebs cycle

1 It is a Krebs cycle intermediate (1); re-cycled from other molecules (1); not used up. (1)　　　　(max 1)
2 reduced NAD is made in the matrix (1); it would move out of the matrix (1); down a concentration gradient (1)
3 could not make enough NAD (1); glycolysis (1); link reaction and Krebs cycle wouldn't occur (1)

129. Oxidative phosphorylation

1 three (1)
2 moves via diffusion (1); from red blood cell through capillary wall (1); from tissue fluid into cells (1); from cytoplasm into mitochondria (1)　　　　　　　　　　(max 3)

130. Anaerobic respiration in eukaryotes

1 covering causes a lack of oxygen (1); respiration becomes anaerobic (1); anaerobic respiration speeds up (1); CO_2 is produced at a faster rate (1)　　　　　　(max 3)
2 to absorb CO_2 released by the yeast (1)

131. Energy values of different respiratory substrates

1 within the mitochondrial matrix (1)
2 The animal would be respiring mainly fats or protein (1); not carbohydrate (1); respiratory quotient is lower for fats and proteins (1); a value of around 0.8 might be expected. (1)
　　　　　　　　　　　　　　　　　　(max 3)

132. Factors affecting respiration

1 ethical considerations/welfare of organism **(1)**; age and size of organism **(1)**; time since last meal **(1)**; diet content **(1)**
(max 2)

2 Without the potassium hydroxide it wouldn't be possible to measure the O_2 used as the CO_2 evolved would replace it. **(1)**

134. Gene mutation

1 substitution mutation **(1)**
and
one base has been replaced with another **(1)**
or
neutral mutation **(1)**
and
GAA and GAG code for the same amino acid. **(1)**
2 Frameshift mutations alter the point at which the genetic code is read. **(1)**
Insertion mutations move the genetic sequence one amino acid to the right. **(1)**
Deletion mutations move the genetic sequence one amino acid the left. **(1)**

135. Gene control

1 Transcription factors control the transcription of genes. **(1)**
Repressor proteins bind to the operator sequence in an operon and prevent the transcription of genes. **(1)**
Proteins are a part of the spliceosome that removes introns from the primary mRNA. **(1)**
2 The repressor protein could no longer bind to lactose. **(1)**
Repressor protein would remain bound to operator. **(1)**
Lac operon genes would not be transcribed. **(1)**
or
The repressor protein would no longer bind to the operator. **(1)**
The lac operon genes could be transcribed **(1)**
even in the absence of lactose. **(1)**

136. Homeobox genes

1 These genes are very important. **(1)**
A mutation would have a big effect / alter body plan. **(1)**
Many other genes would be affected. **(1)**
Mutation likely to be lethal. **(1)**
(max 2)
2 code for transcription factors **(1)**
bind to promoter regions on DNA **(1)**
switch genes on and off / control transcription of genes **(1)**

137. Mitosis and apoptosis

1 allows some cells to die off **(1)**
makes the shape of the limbs / fingers **(1)**
2 Stress / viral infection affects genes in the cell cycle. **(1)**
Cell cycle genes control the cell cycle. **(1)**
Cell no longer goes through mitosis / goes through apoptosis. **(1)**

138. Variation

continuous variation **(1)**
increased melanin in skin **(1)**
environmental influence / UV light **(1)**

139. Inheritance

$\frac{960}{16}$ **(1)**

$= 60$ **(2)** (*Two marks for correct answer; one mark for correct working, if answer incorrect.*)

140. Linkage and epistasis

1 Draw genetic diagram to show inheritance. **(1)**
Possible genotypes of children are $X^H X$ $X^H X$ XY XY. **(1)**
Both daughters are carriers; both sons do not have haemophilia. **(1)**

2 Draw genetic diagram to show inheritance. **(2)**
12 white and 4 purple / ratio 3:1 **(1)**
Only 1 in 4 plants had both dominant alleles. **(1)**

141. Using the chi-squared test

χ^2 value is 4.30. **(2)**
χ^2 value is greater than critical value at $p = 0.05$. **(1)**
Results are significantly different. **(1)**
Reject null hypothesis. **(1)**
(max 4)

142. The evolution of a species

1 allopatric speciation **(1)**
(due to) geographical isolation **(1)**
unable to interbreed **(1)**
different selective pressures on each island **(1)**
2 little genetic variation **(1)**
genetic bottleneck **(1)**
founder effect **(1)**

143. The Hardy–Weinberg principle

$q^2 = \frac{1}{2500} = 0.0004$ or 4×10^{-4} **(1)**

$q = 0.02$ **(1)**
$p = 1 - 0.02 = 0.98$ **(1)**
$p^2 = (0.98)^2 = 0.9604$ **(1)**
$2pq = 1 - 0.9604 - 0.0004 = 0.0392$ or 3.92% **(1)**

144. Artificial selection

Breed together the yellow male dog and the chocolate female; **(1)**
50% of puppies should be yellow. **(1)**
Breed together yellow male dog and one yellow offspring. **(1)**

146. DNA sequencing

Humans, chimpanzees and lemurs share a common ancestor. **(1)**
Humans and chimpanzees shared a common ancestor more recently than humans and lemurs. **(1)**

147. Polymerase chain reaction

1 C and D **(1)**
2 18 cycles **(2)**

148. Gel electrophoresis

Cut the DNA on either side of the gene **(1)**
using restriction enzymes / restriction endonucleases. **(1)**
Run the DNA on the gel. **(1)**
Identify the gene of interest by size (using the DNA ladder). **(1)**
Identify the gene of interest using a DNA probe. **(1)** **(max 4)**

149. Genetic engineering

Extract the human blood clotting factor gene. **(1)**
Cut out the gene using restriction enzymes. **(1)**
Cut open a plasmid / vector using restriction enzymes. **(1)**
Place gene into plasmid / join using DNA ligase. **(1)**
Electroporate goat cells to take up plasmid. **(1)**

150. The ethics of genetic manipulation

1 Yes, because:
animals can produce (named) product in milk **(1)**
greater yield of meat from animals **(1)**
used to research human diseases. **(1)** **(max 2)**
No, because:
can cause pain and suffering to the animal **(1)**
increased susceptibility to disease **(1)**
decreases genetic variation **(1)**
patenting animals treats them like human property. **(1)** **(max 2)**
2 benefit to humans / increased yield / best characteristics **(1)**
increasing numbers of rare / endangered species **(1)**
increased susceptibility to disease **(1)**
decreased genetic variation **(1)**

151. Gene therapy

genetically the same as the individual **(1)**
readily available / might not have kept umbilical cord **(1)**
ethical issues with using embryonic stem cells **(1)**

153. Natural clones

1   Each plant will have the desired characteristics. **(1)**
    Able to produce many plants from one parent. **(1)**
    Plants will all reach same stage of development at the same time. **(1)**
    Quicker than growing the plant from seed. **(1)**
2   not genetically identical **(1)**
    Fraternal twins come from two different fertilised eggs / ova. **(1)**

154. Artificial clones

1   no genetic variation **(1)**
    all susceptible to the same diseases **(1)**
    health problems in cloned animals **(1)**
    Shorter life span in cloned animals **(1)**
    pain and suffering for unsuccessful cloned animals **(1)**
    labour intensive **(1)**
    expensive procedure **(1)**   **(max 5)**
2   Yes, because:
    we could use cloned organs for transplant. **(1)**
    could be used as models in medical research **(1)**
    alternative to IVF for sterile couples **(1)**   **(max 2)**
    No, because:
    cause pain and suffering to unsuccessful human clones **(1)**
    no genetic variation **(1)**
    shorter life span **(1)**
    potential health problems **(1)**
    ethical concerns over human cloning **(1)**   **(max 2)**

155. Microorganisms in biotechnology

1   Low pH could denature enzymes / proteins. **(1)**
    Fungi could die. **(1)**
    decreased yield **(1)**
2   Bacteria would respire anaerobically. **(1)**
    production of ethanol / alcohol **(1)**

156. Aseptic techniques

1   Contamination would result in having to close down the fermenter. **(1)**
    time needed to sterilise the fermenter / production halted during sterilisation time **(1)**
    increased cost / need to replenish nutrient broth and microorganisms **(1)**
    Batch fermentation is stopped at the end of each batch so contamination is not as much of a problem. **(1)**   **(max 3)**
2   Make sure all equipment is sterile. **(1)**
    Make sure agar is sterile before adding bacteria. **(1)**
    Wash hands / wear gloves. **(1)**
    Wear protective clothing / lab coat. **(1)**
    Expose Petri dish to the air as little as possible. **(1)**

157. Growth curves of microorganisms

$28 \times 10\,000 = 280\,000$ per cm^3
$280\,000 \times 10\,cm^3$
$= 2\,800\,000$ bacteria **(3)** (*2 marks for correct working if final answer incorrect*)

158. Immobilised enzymes

Increased temperature could break hydrogen bonds within enzyme / denature enzyme. **(1)**
Active site of enzyme could change shape and no longer be complementary to substrate. **(1)**

Enzyme could no longer bind to substrate / enzyme–substrate complexes could not form. **(1)**
Reaction rate would decrease. **(1)**   **(max 3)**

160. Ecosystems

The number of blue tits would decrease. **(1)**
The blue tits predate on the insects. **(1)**
There will not be enough food for the blue tits and the blue tits will die. **(1)**

161. Biomass transfer

Less energy lost by respiration if movement restricted. **(2)**
More energy converted to biomass / greater NPP. **(2)**

162. Nitrogen cycle

1   *Nitrosomonas* converts ammonium to nitrites. **(1)**
    Nitrobacter converts nitrites to nitrates. **(1)**
    Nitrites and nitrates can be taken up by plants. **(1)**
2   Plants assimilate nitrates to make amino acids for growth. **(1)**
    Too little atmospheric nitrogen is fixed into ammonium. **(1)**
    Too few nitrifying bacteria in the soil. **(1)**
    Too much nitrate is lost from the soil by denitrification. **(1)**
    No plant material is added to the soil due to plant death. **(1)**
                                                        (max 4)

163. Carbon cycle

1   $\dfrac{400-180}{180} \times 100$

    $= 122.2\%$ **(2)** (*1 mark for correct working, if answer incorrect*)
2   It is part of our body structure / amino acids / proteins / cells **(1)**
    use proteins for growth and repair. **(1)**
    It is part of carbohydrates and fats **(1)**
    use carbohydrates and fats for energy. **(1)**

164. Primary succession

1   more nutrients in the soil **(1)**
    increased number of species **(1)**
    increased number of niches **(1)**
    more food available for species **(1)**   **(max 3)**
2   lack of oxygen to the roots **(1)**
    Climax species would die. **(1)**
    Plants tolerant to waterlogged soil / sphagnum moss would outcompete other plants. **(1)**

165. Sampling

1   It is a better representation of the abundance and distribution of organisms. **(1)**
    Can observe how the type of organism changes along the transect. **(1)**
    Can measure how the abiotic and biotic factors change along the transect. **(1)**
2   The population estimate would increase because some of the marked mice would have emigrated. **(2)**
    or
    The population estimate would decrease because there would be fewer mice captured the second time. **(2)**

166. Population sizes

1   (intraspecific / interspecific) competition **(1)**
    predation **(1)**
    disease **(1)**
    food / water availability **(1)**   **(max 3)**
2   interspecific competition **(1)**
    Both species compete for same food source / snowshoe hares. **(1)**
    Only one species can survive. **(1)**
    Wolves outcompete lynx / number of lynx decrease. **(1)**

167. Conservation

1   Preservation is not active or dynamic / or reverse argument **(1)**
    does not allow human intervention / or reverse argument. **(1)**
    Preservation allows nature to take over / conservation
    manages conditions that are best for humans and other
    species. (1)
2   decrease in the number of trees **(1)**
    less timber available **(1)**
    less food availability / shelter for animals **(1)**
    decrease in biodiversity of habitat **(1)**

168. Managing an ecosystem

1   Wildlife attracts tourists. **(1)**
    Tourism threatens wildlife. **(1)**
    income to be made from hunting **(1)**
2   Insect eats native plants. **(1)**
    decrease in food availability for endemic species **(1)**
    no natural predator for insect **(1)**

Notes

Notes

Published by Pearson Education Limited, 80 Strand, London, WC2R 0RL.

www.pearsonschoolsandfecolleges.co.uk

Copies of official specifications for all OCR qualifications may be found on the OCR website: www.ocr.org.uk

Text and illustrations © Pearson Education Limited 2016
Copyedited by Priscilla Goldby
Reviewed by David Barrett
Typeset by Kamae Design
Produced by Out of House Publishing
Illustrated by Tech-Set Ltd, Gateshead
Cover illustration © Miriam Sturdee

The rights of Kayan Parker and Colin Pearson to be identified as authors of this work have been asserted by them in accordance with the Copyright, Designs and Patents Act 1988.

First published 2016

18 17 16
10 9 8 7 6 5 4 3

British Library Cataloguing in Publication Data
A catalogue record for this book is available from the British Library

ISBN 9781447984368

Printed in Slovakia by Neografia

The publisher would like to thank the following for their kind permission to reproduce their photographs:

(Key: b-bottom; c-centre; l-left; r-right; t-top)

123RF.com: 1; **Alamy Images**: Universal Images Group Limited 114; **Colin Pearson**: 121; ©**Crown copyright**, 2013: 76; **Digital Vision**: 168; **Fotolia.com**: Anton Prado PHOTO 20; **Getty Images**: De Agostini Picture Library 71, Ed Reschke 42cl, 49br, 50c, 50bl; **Science Photo Library Ltd**: 37c, ASTRID & HANNS-FRIEDER MICHLER 105, BIOPHOTO ASSOCIATES 50cl, 50b, 62cl, CNRI 4bl, DR JEREMY BURGESS 2tc, 4br, DR KEITH WHEELER 50t, 53bl, 62c (a), 62c (b), 62cr, EYE OF SCIENCE 67tr, GERD GUENTHER 4cr, HEITI PAVES 2cl, National Library of Medicine 67bl, NATURE'S GEOMETRY 67br, NIAID / AMI IMAGES 67tl, ROBERT MARKUS 3b, STEVE GSCHMEISSNER 2tr, 96, WLADIMIR BULGAR 148br; **Shutterstock.com**: Erik Lam 82, gopixa 99, isak55 152bl, kontur-vid 138tl, Roxana Gonzalez 138tr; **SuperStock**: age fotostock 1c

All other images © Pearson Education